Elizabeth Connor, MLS, AHIP

Internet Guide to Pet Health

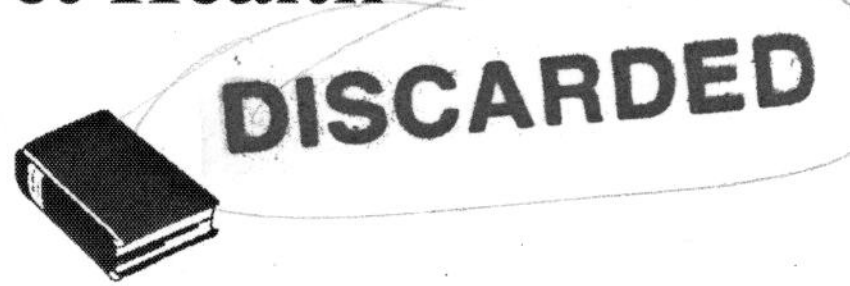

Pre-publication REVIEWS, COMMENTARIES, EVALUATIONS . . .

"This particular reference tool will be of immense value to librarians, veterinarians, graduate students, researchers, zookeepers, laboratory animal raisers, and pet owners who are ever searching for new and updated tools to enhance the health of their animals and pets. . . . Delivers important sources of essential information in the field of veterinary medicine, specifically in the area of pet health. . . . A welcome and timely addition because of the need for reference sources in this particular area and the ever increasing Web sites which have affected how literature searches are conducted on the World Wide Web. This book is comprehensive in nature and is worth any time spent perusing."

Cheryl A. Cumberbatch, MLS,
Library Director, Ross University, School of Veterinary Medicine, West Indies

"A comprehensive overview of high-caliber pet health Web sites. . . . Organized in such a simple and easy-to-follow manner that readers are sure to find what they are looking for faster than they could type keywords in any search engine. . . . Not only a wonderful resource for the pet owner, but also for animal health professionals looking for quality Web resources for their clients."

Tara Tobin Cataldo, MLS,
Biological/Life Sciences Librarian, George A. Smathers Libraries, University of Florida

"Authoritative and comprehensive. . . . Useful. . . . Highlights the most dependable sources of information from educational, organizational, and governmental sites. . . . Recommended for school and public libraries."

Library Journal **(Miriam Tuliao)**

NOTES FOR PROFESSIONAL LIBRARIANS AND LIBRARY USERS

This is an original book title published by The Haworth Press, Taylor & Francis Group. Unless otherwise noted in specific chapters with attribution, materials in this book have not been previously published elsewhere in any format or language.

CONSERVATION AND PRESERVATION NOTES

All books published by The Haworth Press are printed on certified pH neutral, acid-free book grade paper. This paper meets the minimum requirements of American National Standard for Information Sciences-Permanence of Paper for Printed Material, ANSI Z39.48-1984.

DIGITAL OBJECT IDENTIFIER (DOI) LINKING

The Haworth Press is participating in reference linking for elements of our original books. (For more information on reference linking initiatives, please consult the CrossRef Web site at www.crossref.org.) When citing an element of this book such as a chapter, include the element's Digital Object Identifier (DOI) as the last item of the reference. A Digital Object Identifier is a persistent, authoritative, and unique identifier that a publisher assigns to each element of a book. Because of its persistence, DOIs will enable The Haworth Press and other publishers to link to the element referenced, and the link will not break over time. This will be a great resource in scholarly research.

Internet Guide to Pet Health

THE HAWORTH INFORMATION PRESS®
Haworth Internet Medical Guides
M. Sandra Wood, MLS
Editor

The Guide to Complementary and Alternative Medicine on the Internet by Lillian R. Brazin

Internet Guide to Travel Health by Elizabeth Connor

Internet Guide to Food Safety and Security by Elizabeth Connor

Internet Guide to Cosmetic Surgery for Women by M. Sandra Wood

Internet Guide to Anti-Aging and Longevity by Elizabeth Connor

Internet Guide to Herbal Remedies by David J. Owen

Internet Guide to Medical Diets and Nutrition by Lillian R. Brazin

Internet Guide to Cosmetic Surgery for Men by M. Sandra Wood

Internet Resources on Weight Loss and Obesity by Lillian R. Brazin

Internet Guide to Pet Health by Elizabeth Connor

Internet Guide to Pet Health

Elizabeth Connor, MLS, AHIP

For more information on this book or to order, visit
http://www.haworthpress.com/store/product.asp?sku=5675

or call 1-800-HAWORTH (800-429-6784) in the United States and Canada
or (607) 722-5857 outside the United States and Canada

or contact orders@HaworthPress.com

Published by

The Haworth Press, Taylor & Francis Group, 270 Madison Avenue, New York, NY 10016.

Cover design by Marylouise E. Doyle.

Library of Congress Cataloging-in-Publication Data

Connor, Elizabeth, MLS.
Internet guide to pet health / Elizabeth Connor.
p. cm.
Includes bibliographical references and index.
ISBN: 978-0-7890-2977-5 (hard : alk. paper)
ISBN: 978-0-7890-2978-2 (soft : alk. paper)
1. Pets—Computer network resources—Directories. 2. Pets—Diseases—Computer network resources—Directories. 3. Pets—Wounds and injuries—Computer network resources—Directories. 4. Pets—Health—Computer network resources—Directories. 5. Internet addresses—Directories. 6. Web sites—Directories. I. Title.
SF413.C66 2007
025.06'636089—dc22

2007027515

To the many organizations and societies in the greater Charleston area that rescue and care for animals, big and small, especially Pet Helpers, John Ancrum Society for the Prevention of Cruelty to Animals, and Doc Williams Society for the Prevention of Cruelty to Animals.

ABOUT THE AUTHOR

Elizabeth Connor, MLS, AHIP, is Associate Professor of Library Science and Science Liaison at the Daniel Library of The Citadel, the Military College of South Carolina in Charleston, and a distinguished member of the Academy of Health Information Professionals. She has held increasingly responsible library leadership positions at major teaching hospitals and academic medical centers in Maryland, Connecticut, South Carolina, Saudi Arabia, and the Commonwealth of Dominica and now focuses on how undergraduate and graduate students learn science.

Ms. Connor has written numerous peer-reviewed articles; published more than 80 book reviews in several professional library publications; and is the author of *Internet Guide to Travel Health; Internet Guide to Food Safety and Security; Internet Guide to Anti-Aging and Longevity; A Guide to Developing End User Education Programs in Medical Libraries; Planning, Renovating, Expanding, and Constructing Library Facilities in Hospitals, Academic Medical Centers, and Health Organizations; An Introduction to Reference Services in Academic Libraries;* and *An Introduction to Instructional Services in Academic Libraries* (all from Haworth).

CONTENTS

Preface

Each year, more than 27 million dogs and cats are born in the United States. According to the 2003-2004 National Pet Owners Survey conducted by the American Pet Products Manufacturers Association, 39 percent of all American households own at least one dog, 34 percent own at least one cat, and Americans spend $31 billion on products and services for their pets.[1] The Bureau of the Census has estimated that Americans increased their online spending on pet products and services from $400 million in 2003 to $700 million in 2005, and pet expenditures account for 3.2 percent of online spending in 2005.[2]

The health needs of pet animals have become as complex and specialized as those of humans. Pet owners, pet lovers, and veterinarians alike seek authoritative, reliable, and up-to-date information about caring for the health and well-being of pets. Controversies and issues include early spaying or neutering, routine vaccinations, commercial pet foods, declawing of cats, and so forth.

Emerging health issues such as bird flu, West Nile virus, and kidney failure due to tainted pet foods pose serious risks to pets and pet owners alike.

This comprehensive compilation of annotated links will serve as a handy, useful, and easy-to-consult guide for persons who appreciate animals and/or own domestic pets of all kinds, especially cats and dogs. Many diseases and conditions common to household pets are included here but complex topics such as animal breeding or reproduction, pet cloning, pet pharmacy, intensive animal training, animal rights, care of farm animals, insects, or exotic wild animals are not covered. This book contains English-language information predominantly from North American sites, with some links from the United Kingdom, Australia, and some European countries.

Although this guide is intended to be as thorough as possible, use the sites marked with the ☑ symbol to save considerable time and effort when researching these subjects. This symbol highlights unusually comprehensive, well-designed, and/or authoritative sources of information, usually

doi:10.1300/5675_a

from educational, organizational, or governmental sites. Consult the glossary to learn the definitions of words that are unfamiliar or unclear. As scientific knowledge increases exponentially, keep in mind that health care practices that were safe a few years ago can be dangerous today because of research discoveries and changes in economics, environment, and society.

NOTES

1. American Pet Products Manufacturers Association. National Pet Owners Survey, 2003-2004. Available online at: <http://www.appma.org/pubs_survey.asp>. Accessed on May 10, 2007.

2. Table No. 1035. Online Consumer Spending Forecast by Kind of Business: 2003 to 2005. *Statistical Abstract of the U.S.*, 2006.

Chapter 1

Introduction

Pet animals contribute immeasurably to the quality of human life by providing companionship, amusement, physical activity, protection, and much more. Understanding the effects of normal aging, diseases, and health conditions can help pet owners plan for long and healthy lives together. Aging is a natural and lifelong process that starts at birth and can be accelerated by poor eating, lack of exercise, and unsanitary living conditions. Fresh air, healthy food, exercise, loving human interaction, socialization with other animals, and clean living conditions can improve a pet animal's quality of life. It is important for pet owners to schedule regular veterinary checkups for their beloved pets, and consult a veterinarian before embarking on new treatments, diets, or exercise regimens for animals living in the household.

Curious and discerning individuals can use the Internet to find updated and authoritative information on a variety of topics, including pet health. The Internet has the potential to improve health knowledge, and to increase awareness of health risks. Pew Internet & American Life Project reports that "fifty-two million American adults, or 55 percent of those with Internet access, have used the Web to get health or medical information,"[1] and since humans suffer from many of the same diseases and conditions that animals develop, it is entirely possible that some of these people were seeking pet health information.

WEB SITE ADDRESSES EXPLAINED

Hypertext transfer protocol (http) is the set of standards that is used to represent content on the World Wide Web. Although it is no longer necessary to use the http:// prefix when entering site addresses, other prefixes

Internet Guide to Pet Health

doi:10.1300/5675_01

(e.g., ftp and telnet) are understood by browser software to connect to other types of Internet resources. Each Web site address is comprised of distinct and meaningful parts that describe the host computer, directory, and file name:

<protocol://host.domain.suffix/directory/filename>

For example, in the address <http://ohioline.osu.edu/vme-fact/0017.html>, ohioline is the host, osu (Ohio State University) is the domain, edu is the suffix indicating an educational institution, vme-fact is the directory, and 0017 is the file name. In the MedlinePlus site address for pet health <http://www.nlm.nih.gov/medlineplus/petsandpethealth.html>, the domain is the National Library of Medicine (nlm) at the National Institutes of Health (nih), a group of government agencies (.gov), the directory is MedlinePlus, and the file name is petsandpethealth.

The relative ease with which Web sites are designed and content can be uploaded has resulted in many temporary, redesigned, or outdated sites. Dead links result when a site changes file names, alters the site navigation, or stops publishing. If a particular site address no longer functions, delete the /directory, /filename.htm, or /filename.html part of the address, and use the host.domain.suffix parts of the address. After the site loads, use the site's search function (if one exists) to find the specific document or section needed. If this approach does not work, try the wayback machine <http://www.archive.org/web/web.php>, which provides a simple search interface for searching billions of archived Web pages dating back to 1996.

Web site addresses can include organizational or geographic suffixes that are meaningful. Commercial Web sites or sites with a .com suffix at the end of the address can include useful scientific content. Some reputable educational institutions maintain authoritative veterinary information on .com sites with content that is separate from their .edu sites. Some excellent consumer health sites sell brochures and other products, but the sites featured in this guide provide freely available information, including some sites that require completion of a registration process to personalize future site interactions, or association membership for full access to all content on the site.

EVALUATING WEB CONTENT

Consumers of health information should be careful about medical information or advice obtained through the Internet. The currency, accuracy, and source of health-related information are very important factors to consider. Laypersons should be as skeptical and particular as health professionals are when distinguishing between anecdotal information and content derived from authoritative and peer-reviewed sources.

Health on the Net (HON) Foundation <http://www.hon.ch/> is an organization, based in Switzerland, that developed an English and French medical search engine, and a set of standards for evaluating sites with medical content. The HON Code of Conduct rates Web sites according to whether a particular medical site:

- explains qualifications for dispensing advice or developing content
- maintains confidentiality when handling medical information
- attributes information derived from other sources
- indicates when content was modified or revised
- provides detailed contact information for content developers
- identifies sources of funding or sponsorship
- explains the site's use of advertisements or sale of products
- differentiates between original content and promotional content

Health consumers should scrutinize printed or online medical information for relevance and understandability as medical jargon can be unfamiliar and confusing. Depending on individual levels of health knowledge and medical vocabulary, consumers should seek correspondingly simple or complex information. Readers can consult the glossary featured at the end of this work to better understand some of the terms used by persons involved in treating and caring for pet animals.

SEARCH ENGINES/DIRECTORIES

Although government agencies used Internet applications to communicate and share information as early as 1969, the part of the Internet known as the World Wide Web was not developed until 1990. The development of graphical browser tools (Mosaic, Netscape, Internet Explorer, Firefox) and sophisticated search engines/directories have accelerated the growth, development, and acceptance of the Web. The widespread use of these freely available search tools has contributed to the evolution of Web searching as a daily activity, but often search queries yield thousands of marginally relevant results, with some dated or extinct links. Search engines have advantages and drawbacks, and it is worthwhile to learn the features of a few to serve a variety of needs.

A search engine delivers dynamically generated results based on the words typed into the search box. A search directory provides somewhat static groupings of categorized or preselected sites, and tends to be smaller in scope than a search engine. For subjects related to pet health, it may be more productive to focus on several sites with reliable, research-based health content (such as American Veterinary Medical Association, Association for the Prevention of Cruelty to Animals, National Library of Medicine's MedlinePlus, or a veterinary school site) than to enter keywords into a search engine interface and spend hours sorting through links of dubious quality or authority, especially sites that exist chiefly to sell products and services.

Search engines/directories differ greatly in size and in how they are designed, compiled, updated, and organized. Search Engine Watch <http://searchenginewatch.com/>, Search Engine Showdown <http://www.searchengineshowdown.com/>, and Search Engine Blog <http://www.searchengineblog.com/> are excellent sources of information about how specific search tools function and the relative size, advantages, and features of each of the search engines/directories. The following search engines/directories are useful for searching a variety of topics, and were used to locate the links included in this book:

- **Clusty** <http://clusty.com/>: Clusty is a metasearch engine with a mechanism that clusters results from several search engines. A quick search on pet health yields a group of 243 sites clustered into different folders: pet health care (63), insurance (31), supplies (24), nutrition (18), safety (12), food (8), forums (8), and so forth.
- **ExactSeek** <http://www.exactseek.com/>: ExactSeek differs from most search engines in that it delivers fewer but possibly more relevant results based on title metatags and content descriptions. Advanced features include limiting search results to search forms, Boolean logic, and proximity searching. This resource also includes a useful directory of topics ranging from Arts and Humanities to News and Media to Society and Culture.
- **Google** <http://www.google.com/>: Google is an excellent all-purpose and progressive resource for searching or browsing content including publicly accessible Web sites, news group messages, images,

shopping price comparisons (Froogle), scholarly information (Google Scholar), full-text content from books (Google Book Search), and much more. The subject categories in the Google Directory <http://directory.google.com/> can be browsed and searched. Features include the caching of old pages, linking to similar content, and narrowing results to specific universities, type of sites, type of file formats, and other variables.
- **Yahoo!** <http://www.yahoo.com/>: The advantages of Yahoo! as a search directory include compilation by humans, organization, speed, and ease of use. If Yahoo! exhausts its directory index, the search query is referred to its search engine. Features include image, video, audio, news, local, and product searching.

Several search engines and directories focus on specific areas of interest such as government or medicine. These specialized resources may be useful to research topics related to pets and their health needs:

- **USA.gov** <http://www.usa.gov/>: USA.gov is the U.S. government's Web portal, an aggregated interface that can be used to search for content available on federal and state government sites. USA.gov for Science <http://www.science.gov/> is a portal subset that focuses on science, specifically authoritative information available from twelve U.S. government agencies, including the Departments of Agriculture, Commerce, Defense, Energy, Education, Health and Human Services, Interior, Environmental Protection Agency, NASA, National Science Foundation, and the Government Printing Office. In addition to searching these agencies and their databases, it is possible to browse topic areas such as agriculture and food; applied science and technology; biology and nature; environment and environmental quality; and health and medicine, among others.
- **SearchEdu and SearchGov** <http://searchedu.com> <http://searchgov.com>: Despite the .com suffix on each of these site addresses, the interfaces retrieve results, which are limited to education (.edu), and government (.gov) sites respectively, making it easier to focus on authoritative content especially from veterinary school, health agency, or agricultural agency sites.

- **Scirus** <http://www.scirus.com/>: As of September 2007, Scirus "searches over 450 million science-specific Web pages," by focusing on scientific, peer-reviewed journal content including e-prints (electronic article content), preprints (articles available before release of print publication), citations, and full-text literature. Nonscientific sites are filtered out of the search results. Results can be restricted to specific years, journals, authors, and subjects within science, technology, and medicine.

WEB 2.0 (DISCUSSION GROUPS, RSS FEEDS, AND BLOGS)

Web 2.0 is a social phenomenon that refers to the use of various technological "tools to create, aggregate, and share dynamic content in ways that are more creative and interactive than transactions previously conducted on the Internet."[2] Tools of particular interest to pet owners include online communities, Web logs (blogs),[3] and RSS feeds for persons with specialized interests. Researchers associated with the Pew Internet & American Life Project have estimated that as many as 27 percent of Americans who go online read blogs, which fulfill unmet needs to communicate according to Friedman.[4] Bloggers share information through e-mail or Web-based interfaces. With MetaFilter (MeFi), multiple persons post content to the same Web-based log in a sequential, almost conversational way. RSS (really simple syndication) pipes Web content to other sites. For example, a site can display news headlines, stock quotes, horoscopes, or other frequently updated content from another site or an individual can access syndicated content through the use of RSS readers or aggregators. While these tools have revolutionized the dissemination of ideas and challenged traditional ways of reporting news information, some of the resulting content is inaccurate, unscientific, and unsubstantiated. Anecdotal information about health conditions and their treatments may be more harmful than helpful, and consumers should be discerning and sometimes skeptical readers.

STAYING WELL-INFORMED AND KEEPING PETS HEALTHY

To keep pets healthy and acting energetic and youthful, readers should learn as much as possible about pet health and well-being. The quality and length of a pet's life can be attributed to a combination of genetics, healthy living, moderate exercise, balanced diet, regular health care, and humane treatment. Pet owners can improve their knowledge of pet health by focusing on authoritative sources of information, and avoiding unsubstantiated consumer health information.

UNSUBSTANTIATED INFORMATION

A segment of American society thrives on the dissemination of urban legends, meaning, recent but largely untrue stories that are spread by word of mouth (and lately, by e-mail, online discussion groups, blogs, and Web sites) and can reach epic proportions. Health consumers should be aware of potential fraud and quackery when seeking information about pet health treatments and products. Fictitious examples related to pets include the dangers of floor cleaners, some cases of transplantation of animal tissues to humans, pets being cooked in microwave ovens, and more. Harmful and harmless elixirs, snake oil, and potions have existed since the beginning of time, but discerning consumers can consult a number of resources to understand and dispel unsubstantiated health claims that may harm themselves or their pets.

Quackwatch <http://www.quackwatch.com/> is a corporation developed and maintained by Stephen Barrett, MD, and its "purpose is to combat health-related frauds, myths, fads, and fallacies." National Council Against Health Fraud (NCAHF) <http://www.ncahf.org/> is a useful resource for determining whether a particular health treatment is fraudulent or not. Much of the NCAHF site focuses on humans but some of the content deals with illegal veterinary pharmacies, alternative treatments for animals, and the like. Another useful source of reliable information about veterinary treatments is Veterinarywatch <http://www.veterinarywatch.com/>.

To help determine whether an anecdote is truth, folklore, or a hoax, consult sites that collect and report on such stories including Urban Legends Reference <http://www.snopes.com/>, Urban Legends and Folklore

<http://urbanlegends.about.com/>, or Museum of Hoaxes <http://www.museumofhoaxes.com/>. These sites can help to dispel various falsehoods of interest and concern to pet owners.

SOURCES OF AUTHORITATIVE INFORMATION

Some government and university sites that focus on aspects of pet health include U.S. Department of Agriculture, U.S. Food and Drug Administration, U.S. Administration on Aging, land grant universities with agricultural and veterinary programs, and resources developed by the National Library of Medicine. The United States maintains two national libraries that provide services and resources that support scientific research that is crucial to veterinary research and tangential to pet health. The National Agricultural Library (NAL) <http://www.nal.usda.gov/> in Beltsville, Maryland strives to improve life by providing access to agricultural information, most notably through its collections, services, and resources such as AGRICOLA, a bibliographic database that covers many aspects of agriculture, animal science, food and nutrition, water quality, and related subjects. The National Library of Medicine (NLM) <http://www.nlm.nih.gov/> in Bethesda, Maryland is the largest biomedical library in the world and produces an array of bibliographic and full-text resources including MEDLINE and MedlinePlus.

NOTES

1. *PIP Health Report*. Pew Internet & American Life Project. Available online at: <http://www.pewinternet.org/pdfs/PIP_Health_Report_July_2003.pdf> Accessed on May 10, 2007.
2. Connor, Elizabeth. "Medical Librarian 2.0." *Medical Reference Services Quarterly* 2007; 26(1): 1-15.
3. *The State of Blogging*. Pew Internet & American Life Project, 2005. Available online at: <http://www.pewinternet.org/pdfs/PIP_blogging_data.pdf>. Accessed on May 10, 2007.
4. Friedman, Thomas L. (2005). *The World Is Flat: A Brief History of the Twenty-First Century*. New York: Farrar, Straus, and Giroux.

Chapter 2

General Pet Health Sites

INTRODUCTION

Many pet health sites feature specific and useful information about different types of pets, health concerns, and care. Use these broad pet health sites to research both routine topics and unusual conditions.

AltVetMed
<http://altvetmed.org/>

AltVetMed is a site that was developed by two veterinarians and focuses on complementary and alternative medical treatment for animals including articles, news, discussion forums, organizations, meetings, publications, suppliers, and more.

☑ Ani-Med
<http://www.animed.org/>

Ani-Med, a site which is sponsored by the American Society for the Prevention of Cruelty to Animals (ASPCA), has two different sections: (1) information about dogs and cats, and (2) information about birds, fish, rabbits, ferrets, guinea pigs, hamsters, mice, gerbils, and rats. Site features include information about choosing a cat or dog, health problems, behavior problems (aggression, soiling, separation anxiety, training, etc.), and responsible pet ownership (spaying/neutering, pet loss, pet insurance, reporting animal cruelty, and other topics). The information about small

Internet Guide to Pet Health

doi:10.1300/5675_02

pets includes behavioral problems, daily care checklist, household dangers, toys, traveling, and more.

☑ Animal Health Care <http://www.animalhealthcare.ca/>

The Canadian Veterinary Medical Association developed this site with pet health content in English and French languages. Site features include separate sections for cats, dogs, farm animals, and other animals (birds, rabbits, fish, and pocket pets). Animal health guides cover specific topics such as vaccinations for dogs or cats, dog bite prevention, choosing dog collars, zoonotic diseases, dealing with pet loss, hearing-impaired dogs, toxic houseplants, and much more (Figure 2.1).

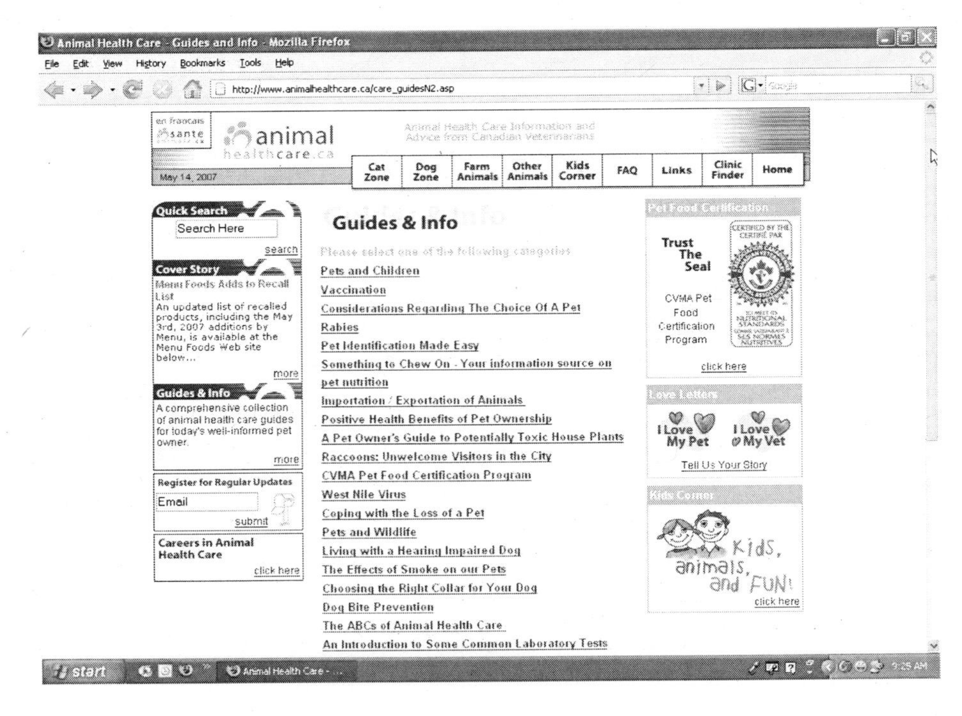

FIGURE 2.1. Animal Health Care
<http://www.animalhealthcare.ca/>
Reprinted with permission.

Animal PetDoctor
<http://animalpetdoctor.homestead.com/>

Roger Ross is a veterinarian who practices animal medicine in Seneca, South Carolina. This site includes easy-to-understand and useful information about the treatment of various diseases and conditions that are commonly seen in cats and dogs. Topics include "injuries, cancer, diarrhea, heart & kidney disease, fractures, bladder infections, leukemia, parvo, diabetes, and much more." The site also discusses vaccinations; diet and nutrition; treatment of fleas, ticks, worms, and mange; responsible pet ownership; caring for stray animals; special bonds shared by humans and pet animals; and more.

☑ **Animalhealthchannel**
<http://www.animalhealthchannel.com/>

Animalhealthchannel "provides comprehensive, trustworthy information about conditions and diseases that affect animals, as well as pet care and training information," and its content is "developed and monitored by board-certified veterinarians." This site is organized into major parts such as conditions (allergy, distemper, emergencies, feline infectious peritonitis, feline immunodeficiency virus, feline leukemia virus, neutering, rabies, spaying, worms, and more); behavior (aggression, aversion to using a litter box, introducing a new pet, introducing a baby to a pet, and more); and resources (bibliography, finding a veterinarian, links to external sites, video clips, and more).

☑ **Care for Pets**
<http://www.avma.org/care4pets/default.htm>

The American Veterinary Medical Association (AVMA) site's information about pet care focuses on several main areas of interest: pet health, pet loss, selecting a pet, animal safety, information about veterinarians, and activities for children interested in learning more about pets and pet care.

☑ **HealthyPet**
<http://www.healthypet.com/>

HealthyPet is a site that was developed and is maintained by the American Animal Hospital Association (AAHA). Site features include information about animal hospitals, a pet care library with articles about dogs, cats, and "exotic" animals (birds, hamsters, iguanas, rabbits, etc.). Topics covered include common health problems, illnesses, diseases, nutrition, preventive care, and more.

☑ Intute—Veterinary
<http://www.intute.ac.uk/healthandlifesciences/veterinary/>

Formerly known as VetGate, this site describes itself as "a free online service providing you with access to the very best web resources for education and research, evaluated and selected by a network of subject specialists." The site features include a tutorial for learning about the Internet, links to veterinary jobs, and hot topics (multiple sclerosis, bird influenza, food safety, etc.).

☑ Martindale's The Virtual Veterinary Center
<http://www.martindalecenter.com/Vet.html>

Jim Martindale's site is organized into several major parts such as veterinary journals; overview of veterinary topics such as laboratory diagnosis; treatment protocols; chemical safety guidelines; veterinary school curricula and online courses; clinical subjects; drugs used to treat animal diseases; a variety of animal diseases with different causes (bacterial, parasitic, viral, etc.); physical diagnosis; and vaccinations.

Office International des Epizooties (OIE)
<http://www.oie.int/>

OIE was established in 1924 by the League of Nations, the forerunner of the United Nations, and concerns itself with world animal health including disease data, health standards, publications, and more.

People's Dispensary for Sick Animals (PDSA)
<http://www.pdsa.org.uk/>

PDSA is a U.K.-based veterinary charity. The charity's site includes information for owners of dogs, cats, rabbits, gerbils, guinea pigs, hamsters, rats, mice, birds, ferrets, and fish.

The Pet Center
<http://www.thepetcenter.com/>

PetFoodDirect.com owns and operates this site with content developed by practicing veterinarians. This site functions as a virtual animal hospital with content and images that explain physical examination, radiology, surgery, and drug therapy for diseases and conditions common to dogs and cats. Other site features include information about pet care, pet photographs, and pet supplies.

Pet Columns
<http://www.cvm.uiuc.edu/petcolumns/>

Experts from the University of Illinois Urbana-Champaign's College of Veterinary Medicine explain topics about birds, cats, dogs, horses, livestock, reptiles, and wildlife. Past columns have discussed egg laying in birds, pet dental care, itchy skin in dogs, iguana care, and more.

☑ PetEducation.com
<http://www.peteducation.com/>

Veterinarians and pet product suppliers, Race Foster and Marty Smith, developed this excellent resource for pet owners of dogs, cats, fish, birds, ferrets, reptiles, and other small pets (gerbils, hamsters, mice, rats, rabbits, guinea pigs, chinchillas, hedgehogs, sugar gliders). Information includes first aid and emergencies, diagnosis and treatment specific to an animal or generalized to several kinds, dictionary, drug information, alternative medicine options, and help hotlines. Other site features include articles related to specific species (setting up an aquarium, ferrets that bite, wool sucking in cats, etc.), video clips (cleaning a dog's ears, cleaning a dog's teeth), and more (Figures 2.2 and 2.3).

Pet Health—Novartis Animal Health
<http://www.ah.novartis.com/pet_health/en/index.shtml>

Novartis Animal Health manufactures a wide variety of products that benefit pet animals and farm animals, including treatments for heartworm, fleas, and roundworms. Site information about pet health focuses on dogs and cats. Information about dogs includes information about fleas, parasitic worms, separation anxiety, atopic dermatitis, and heart diseases. Information about cats includes fleas, parasitic worms, and kidney diseases.

FIGURE 2.2. PetEducation.com
<http://www.peteducation.com/>
Reprinted with permission.

Pethealthcare
<http://www.pethealthcare.co.uk/>

Pethealthcare is a site that focuses on dog care and breeds, cat care and breeds, choosing first pets, latest news, veterinarians, and rescue centers in the United Kingdom, and pet health or behavior questions answered by experts. The site features a personal file area for keeping vaccination and other health records for pets, which is accessible by personal password.

☑ Pet Health Information
<http://www.animalhospitals-usa.com/>

The site of Animal Hospitals USA features useful information about illnesses and injuries in "dogs, cats, birds, guinea pigs, gerbils, ferrets,

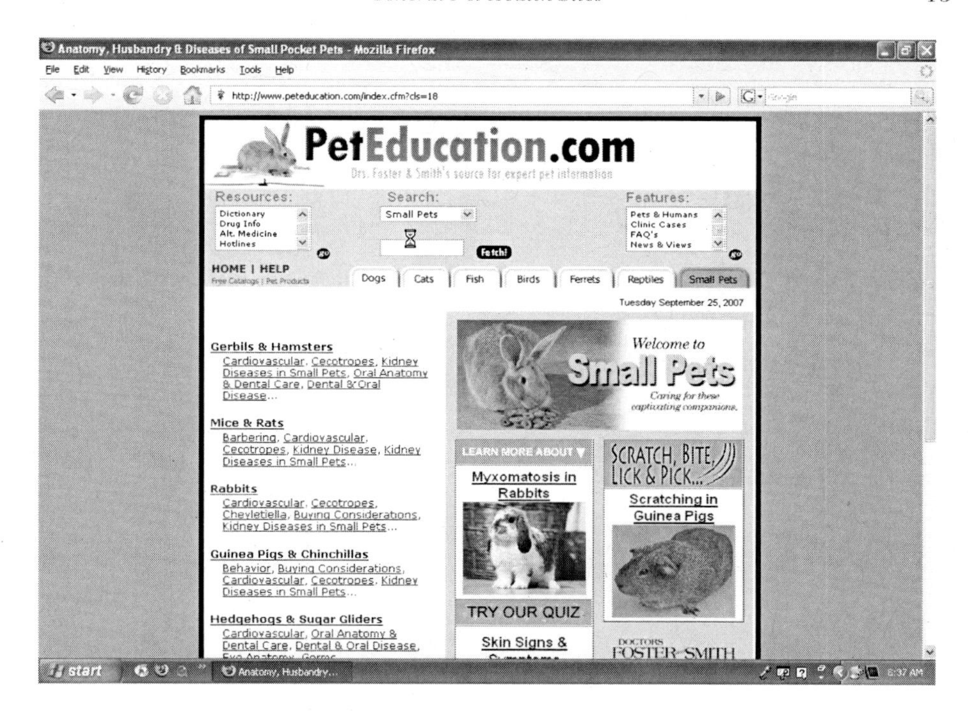

FIGURE 2.3. PetEducation.com—Small Pets
<http://www.peteducation.com/index.cfm?cls=18>
Reprinted with permission.

prairie dogs, and reptiles." For example, the section about reptiles includes images and text related to iguanas, lizards, snakes, water turtles, box turtles, and desert tortoises. Typical sections feature information about health, emergencies, routine care, vaccinations, grooming, training, and breed characteristics.

Pet Health Topics <http://www.vetmed.wsu.edu/ClientEd/>

Washington State University's College of Veterinary Medicine site includes pet health topics (bunny care, adopting a pet, determining the sex of a kitten, etc.), anatomy (text and images related to dog or cat organ systems), and definitions and abbreviations of common veterinary terms. Pet health topics range alphabetically from acupuncture to vomiting, or thematically (procedures, common problems, diseases, miscellaneous health topics).

Pet Parents
<http://www.petparents.com/>

Bayer, the well-known aspirin manufacturer, also manufactures animal health products such as treatments against fleas, ticks, and mosquitoes. Bayer's Pet Parents site includes useful information about pet names, pet care, traveling with pets, electronic newsletter, and a choice of cat or dog electronic cards, selection of pet-related news, and a "family" album of pet pictures. The site also includes an "Ask a Vet" feature that allows site visitors to use a site-based form to ask questions, and to review an archive of past questions and answers.

☑ PetStyle
<http://www.petstyle.com/>

PetStyle is a site that provides useful information about cats and dogs. The site features include information about pet-friendly accommodations, breed profiles, and breeders of cats and dogs, pet forum topics, sources of pet products, and more.

☑ Pet Web Site
<http://www.petwebsite.com/>

Pet Web Site, based in Oxfordshire, United Kingdom, provides useful information related to cats, dog, rabbits, gerbils, guinea pigs, horses, mice, and rats. Topics on the site include pet ownership, pet loss, routine care, training, and more. The information about each type of pet covers supplies and equipment, feeding, handling, grooming, training, and health.

PetPlace.com
<http://www.petplace.com/>

Veterinarian Jon Rappaport's site features separate pet centers for cats, dogs, birds, horses, reptiles, fish, and small mammals (guinea pigs, pot-bellied pigs, rabbits); drug library; medical center; veterinarian locator; and information about pet insurance.

Pets Health
<http://www.healthcentral.com/channel/408/1704.html>

HealthCentral.com is a site that focuses on women's health, men's health, diet, kid's health, senior health, and a number of centers, including one about pet health. The pet health information includes encyclopedic entries about various subjects (fleas, rabies, etc.) and advice from physician-broadcaster Dr. Dean Edell.

☑ Pets and Pet Health
<http://www.nlm.nih.gov/medlineplus/petsandpethealth.html>

MedlinePlus, which is produced by the U.S. National Library of Medicine (NLM), provides access to authoritative, full-text information on more than 650 diseases and conditions of potential interest to consumers and health professionals alike. The information about pets includes overviews, treatment, prevention/screening, nutrition, specific conditions (arthritis, West Nile virus, cancer, distemper, etc.), research, directories, organizations, law and policy, children, and pregnant women (Figure 2.4).

Ponsonby Veterinary Centre
<http://www.petvet.co.nz/>

Ponsonby Veterinary Centre is located in Lynn, Auckland, New Zealand. Its site features health topics about cats and dogs (vaccinations, intestinal worms, spaying/neutering, fleas, dental care, and more); small pets (feeding rabbits, guinea pigs, mice, or rats; neutering rabbits; rabbit viral hemorrhagic disease, dental care for rabbits or rodents); and a free newsletter. Recent newsletter topics have included senior pets, microchipping regulations, fireworks phobias, and hairballs.

Preventative Medicine
<http://www.wcvh.com.au/wcvh.php?sec=52>

Weston Creek Veterinary Hospital in Waramanga, Australia developed part of its site to explain preventive measures (vaccinations, tick/flea/heartworm medications, spaying/neutering, dental care, microchipping, coat

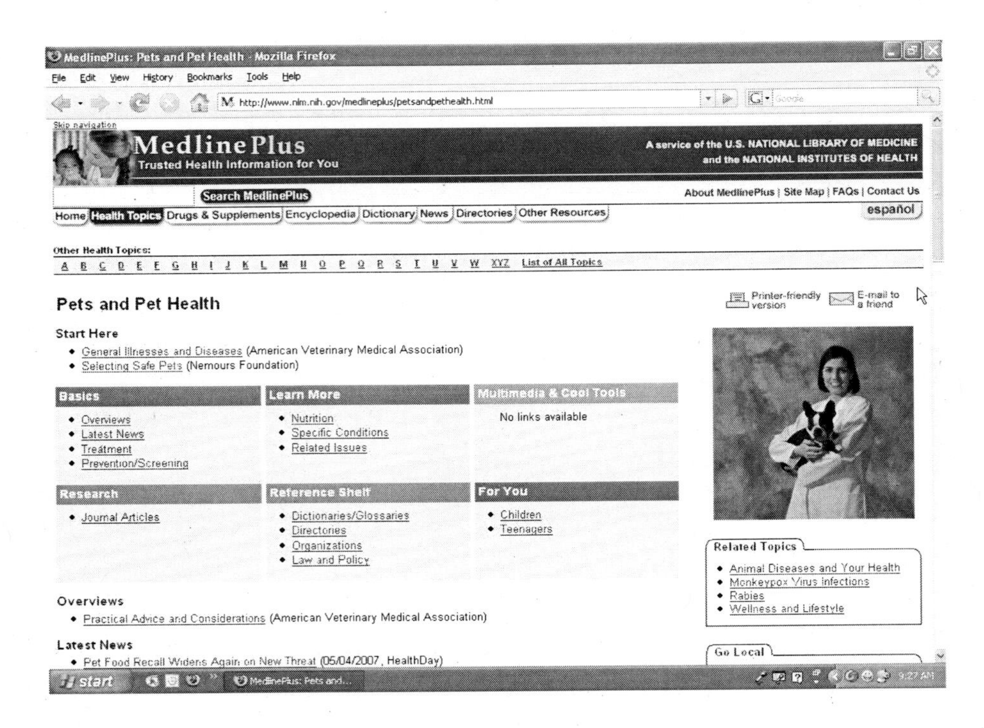

FIGURE 2.4. Pets and Pet Health
<http://www.nlm.nih.gov/medlineplus/petsandpethealth.html>

bathing, and brushing); diet and nutrition; exercise; and other topics of interest to pet owners.

☑ VetCentric.com <http://www.vetcentric.com/>

VetCentric is a site that focuses on the health and behavior of dogs, cats, horses, birds, fish, ferrets, rabbits, "pocket pets" (tarantulas, hamsters, sugar gliders, guinea pigs, mice, chinchillas, etc.), reptiles, farm animals, and wildlife. Site features include an encyclopedia of common and not so common diseases, glossary of medical terms, and frequently asked pet health questions.

☑ VeterinaryPartner.com <http://www.veterinarypartner.com/>

The Veterinary Information Network (VIN) developed this site to provide "veterinarians, veterinary students, and the veterinary industry with the information and tools they need to address the demands of modern veterinary practice." Site features include pet care, behavior, health (grooming, dental care, nutrition, etc.), diseases, and drug information specific to dogs, cats, reptiles, and small mammals (rabbits, ferrets, chinchillas, gerbils, hamsters, mice, rats, guinea pigs). A number of veterinary sites use information from VIN to provide pet health content.

☑ VetMedCenter <http://www.vetmedcenter.com/>

This resource is divided into two tracks: one is for veterinarians, and the other is for pet owners. The information for pet owners focuses on pet wellness, behavior, nutrition, first aid, and the challenges and benefits of living with pets. The site features include conducting wellness assessments on dogs or cats, medical news on a variety of topics (feline infectious peritonitis, food allergies, diagnosing fungal infections, etc.), cat breed center, dog breed center, and more.

Vetservice.co.nz—Pet Health Center <http://www.vetservice.co.nz/content/pet_index.php>

Vetservice.co.nz is a group of 106 New Zealand veterinary practices whose mission is "to lead the veterinary industry through the development of commercial excellence and the delivery of innovative products and services." Their site includes a comprehensive section for owners of cats, dogs, horses, and small pets. The highlights of this site include a pet magazine (*PAWS for Thought*), articles written by practicing veterinarians, and featured pets. Some site content is restricted to veterinarians.

INTERACTIVE TOOLS

Interactive tools include calculators, databases, and games that provide dynamic information. Use these sites to customize the content needed to keep your pet animal happy and healthy.

Animal Webcam Locator
<http://www.webcamlocator.com/animals/animals_index.html>

This site organizes links to various Web cameras (cams) that provide entertaining real-time glimpses into the daily lives of domestic and wild animals including alligators, alpacas, fish in aquariums, badgers, bears, birds, cats, chinchillas, dogs, eagles, elephants, falcons, ferrets, frogs, giraffes, gorillas, horses, rodents, waterfowl, wolves, and animals in zoos.

☑ **Animaland.com**
<http://www.animaland.org/>

ASPCA's Animaland features family-friendly content related to pet care (dog, cat, bird, fish, rabbit, gerbil, hamster, guinea pig, mouse, rat, etc.), animal encyclopedia, games and cartoons, humane education, real issues (use of animals in product testing, pet loss, reporting animal cruelty, animal overpopulation, etc.), career center, and more (Figures 2.5 and 2.6).

Care for Animals
<http://www.avma.org/careforanimals/animatedjourneys/animatedfl.asp>

The American Veterinary Medical Association (AVMA) site's section about pet animal care includes animated clips for selecting a pet and pet health, seasonal tips, and animal-related activities for children.

Consultant
<http://www.vet.cornell.edu/consultant/consult.asp>

Cornell University's College of Veterinary Medicine developed this diagnostic support system that can be searched by diagnosis or sign. For example, by choosing feline as the animal species and entering "vomiting" as a sign, the database lists "Vomiting or regurgitation, emesis" and "Hematemesis, vomiting blood" as possible diagnoses. A list of signs organized according to body system (acoustic to urinary) ranges from deafness or tachycardia to diarrhea or tremor.

☑ **DogAge Test**
<http://www.dogage.com/reg/regassess.aspx>

The developers of RealAge, a resource that helps humans calculate their "real" physiological age based on a number of lifestyle factors, developed

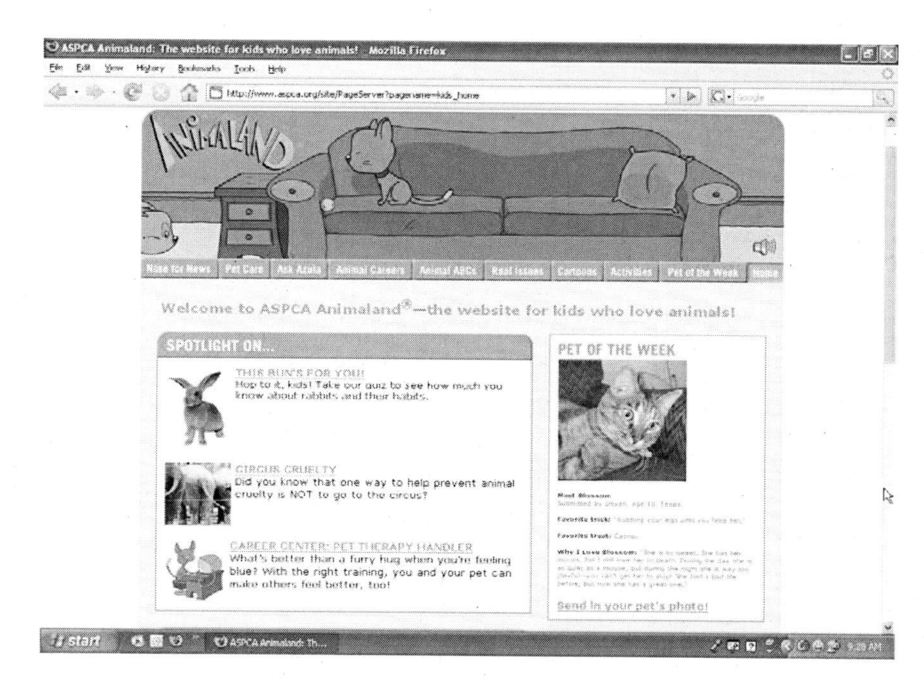

FIGURE 2.5. Animaland
<http://www.animaland.org/>
Copyright © 2007 The American Society for the Prevention of Cruelty to Animals (ASPCA). Reprinted with permission of the ASPCA. All Rights Reserved.

this site to help people calculate the "real" ages of their pets by applying this same principle to dogs. Site features include testing a dog's age based on a number of factors, canine care (preventive measures, health conditions, food and nutrition, exercise, safety), and much more.

Find a Vet
<http://www.findavet.org.uk/>

The Royal College of Veterinary Surgeons in the United Kingdom developed this resource for persons interested in locating veterinary surgeons in the United Kingdom. The database is searchable by name of the veterinary practice, town, postal region, or type of practice (pets, farm animals, exotic animals, etc.).

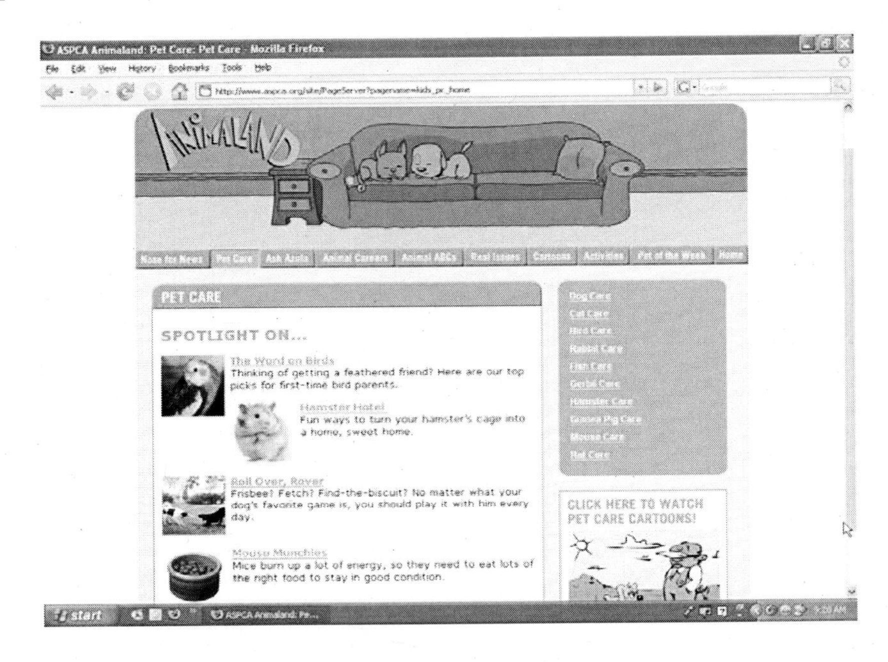

FIGURE 2.6. Animaland Pet Care Guide
<http://www.animaland.org/framesets/abcs_frameset.asp>

Pet of the Day <http://petoftheday.com/>

Karen and Paul Watts have developed this well-designed site that features daily color images and narratives about a pet (usually birds, dogs, or cats), pet talk discussion boards, results for daily poll questions, and more.

Pet Loss Hotline—Cornell University <http://weww.vet.cornell.edu/Org/PetLoss/>

Cornell University's Pet Loss Support Hotline is staffed by volunteer veterinary students. The site includes information about the hotline service, pet euthanasia, other pet loss hotlines staffed by veterinary students, stages of grieving for deceased pets, and links to other sites related to pet loss.

Pet Loss Hotline—Washington State University
<http://www.vetmed.wsu.edu/PLHL/>

Washington State University's College of Veterinary Medicine site includes pet health topics (bunny care, adopting a pet, determining the sex of a kitten, etc.), anatomy (text and images related to dog or cat organ systems), and definitions and abbreviations of common veterinary terms. The pet loss information stresses that it is all right "to love and miss your pet," and includes a hotline service staffed by veterinary students.

Pets911.com
<http://www.pets911.com/>

Pets911 features a toll-free bilingual hotline (1-888-PETS-911) and Web site with useful information about adopting a pet; finding a local shelter, rescue organization, veterinarian, or emergency clinic by ZIP code; and health information for cats and dogs.

☑ Toxic Plant Database
<http://www.library.uiuc.edu/vex/toxic/intro.htm>

Retired veterinary librarian Mitsuko Williams of the University of Illinois Urbana-Champaign's College of Veterinary Medicine developed this database of plants that are known to be toxic to animals. The entries are listed by common names and scientific names, and include a color image of each plant.

ORGANIZATIONS AND SOCIETIES

During the mid to late nineteenth century, Americans became increasingly concerned about health, diet, exercise, and cruelty toward animals. The organizations that fostered humane treatment of animals resulted in similar organizations and social reforms needed to prevent abuse and neglect of children in the United States.[1]

American Animal Hospital Association (AAHA)
<http://www.aahanet.org/>

AAHA strives to promote healthy practices that result in healthier pet animals. Site features include information about accreditation standards for veterinary practices, education programs for veterinary personnel, some site information, and full-text publications restricted to members, state veterinary organization links, and more.

American Association of Feline Practitioners (AAFP)
<http://www.aafponline.org/>

AAFP is a professional organization of veterinarians who practice feline medicine and surgery. Site features include practice guidelines (feline behavior, retroviral testing, feline vaccines, diseases transmitted from cats to humans, antimicrobial therapy, and care of senior cats) and resources limited to members.

American College of Veterinary Internal Medicine (ACVIM)
<http://www.acvim.org/>

ACVIM describes itself as "the official organization of the veterinary specialties of small animal internal medicine, large animal internal medicine, cardiology, neurology, and oncology." Site features of note to pet owners include telling poignant animal survivor stories and explaining the significance of board-certified veterinarians.

The American Humane Association
<http://www.americanhumane.org/>

This association is dedicated to the prevention of "cruelty, abuse, neglect, and exploitation of children and animals." Site features include information about pet adoption, pet care, issues such as spaying and neutering, prevention of cruelty or harm to animals during the production of motion pictures and television programs, animal-friendly legislation, public education and awareness activities, animal shelters, disaster relief, and more.

American Kennel Club (AKC)
<http://www.akc.org/>

AKC focuses on purebred dogs, breeds, breeders, owners, and events. Site features relevant to pet health include canine legislation, dog training, establishing community dog parks, importance of nutrition, and more.

☑ American Society for the Prevention of Cruelty to Animals (ASPCA) <http://www.aspca.org/>

ASPCA is a nonprofit organization that focuses on preventing animal cruelty through educational and awareness efforts. The ASPCA site includes many features of interest to owners of dogs, cats, horses, rabbits, guinea pigs, hamsters, gerbils, rats, mice, and more. Features include a toll-free poison control center that charges a $50 consulting fee, a series of articles called *Companion Lines,* pet check video clips, information about pet care and nutrition, traveling with pets, bioterrorism, emergency preparedness, and more (Figure 2.7).

FIGURE 2.7. American Society for the Prevention of Cruelty to Animals (ASPCA) <http://www.aspca.org/>

☑ American Veterinary Medical Association (AVMA) <http://www.avma.org/>

AVMA "is a not-for-profit association representing more than 72,000 veterinarians working in private and corporate practice, government, industry, academia, and uniformed services." Site features include sections about animal care, veterinary news, animal welfare, vaccination, disaster preparedness, zoonotic agents of bioterrorism, and veterinary publications, with some content restricted to members of the organization. The site also includes statistics related to various pet demographics and numbers of veterinarians, veterinary practices, and more. The information about caring for animals includes animated clips about the joys of pet ownership, caring for pets during the different seasons of the year, play activities for children, dental health, getting a new kitten, and getting a new puppy.

Animal Health Information Specialists (UK and Ireland) <http://www.ahis.org/>

Animal Health Information Specialists (AHIS) is "a group of library and information professionals working in the area of animal health," in the United Kingdom and Ireland. The organization convenes an annual meeting, publishes a newsletter twice a year, and hosts an e-mail-based discussion list. The organization's Web site features external links to veterinary schools and to animal health resources.

Animal Health Institute <http://www.ahi.org/>

Animal Health Institute represents animal health product manufacturers and is based in Washington, D.C. The site features include information about keeping pets healthy, healthy food supplies, use of antibiotics, government regulations of drugs used with animals raised for food, and a useful glossary of terms.

Animal Health Trust (AHT) <http://www.aht.org.uk/>

AHT is a registered charity and research institute, whose patron is the Queen of England, and president is Anne, the Princess Royal. AHT focuses on the health and well-being of cats, dogs, and horses through research and education. Site features include news; pet information leaflets on a variety of topics including deafness, hip dysplasia, eye diseases, etc.; e-mail list about the trust's activities; and more. Some content is restricted to AHT staff.

☑ Animal Welfare Information Center (AWIC) <http://awic.nal.usda.gov/>

AWIC at the National Agricultural Library of the U.S. Department of Agriculture (USDA) provides "information for improved animal care and use in research, teaching, and testing." Site features include hot topics, news, government and legal resources, publications, databases, and information about laboratory, farm, companion, and wild animals (Figure 2.8).

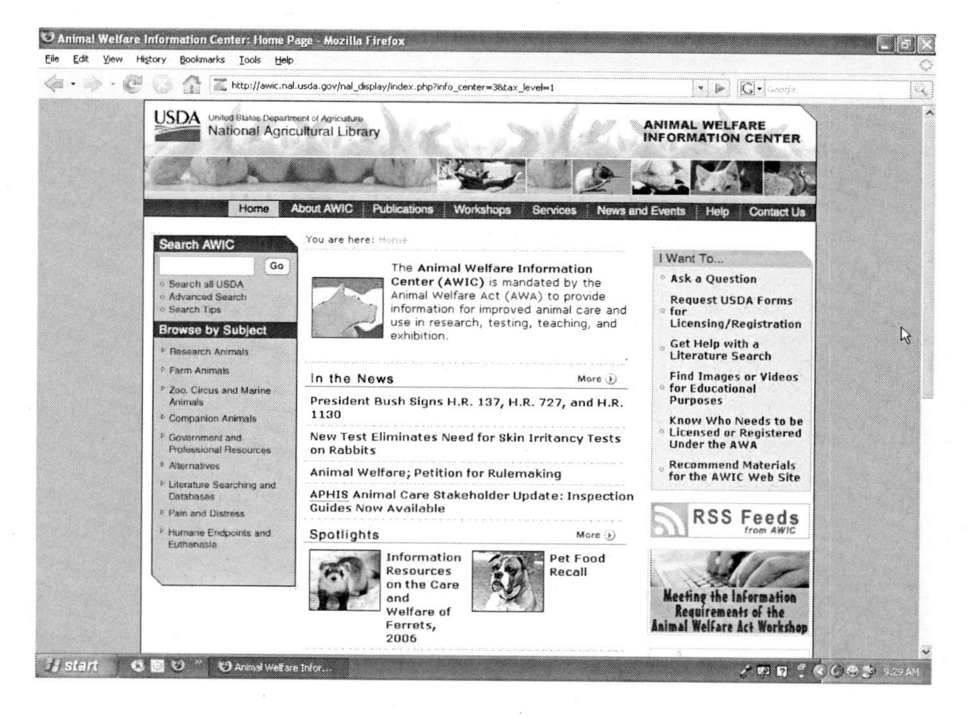

FIGURE 2.8. Animal Welfare Information Center (AWIC) <http://awic.nal.usda.gov>

British Veterinary Association (BVA)
<http://www.bva.co.uk/>

BVA differs from the Royal College of Veterinary Surgeons (RCVS) in that the former is a voluntary organization, and the latter is statutory in nature. The BVA site includes a number of features that are restricted to members but the public can access information about policy formulation, lobbying activities, inherited diseases in dogs, and links to external sites.

Canadian Animal Health Institute (Institut Canadien de la Sante Animale)
<http://www.cahi-icsa.ca/>

Canadian Animal Health Institute is "the trade association representing the developers, manufacturers, and distributors of animal pharmaceuticals, biologicals, feed additives, and animal pesticides in Canada." Site features of interest to owners of pet animals include understanding vaccines, handling the loss of a pet, and risks to pets from West Nile virus.

Canadian Veterinary Medical Association (CVMA)
<http://canadianveterinarians.net/>

CVMA "is the national body serving and representing the interests of the veterinary profession in Canada." Content includes news and events, directory of Canadian veterinarians and veterinary clinics, fact sheets, journal article abstracts, and more. Some features on the CVMA's site are restricted to members of the organization (practicing veterinarians and veterinary students).

☑ Center for Veterinary Medicine (CVM)
<http://www.fda.gov/cvm/default.html>

The U.S. Food and Drug Administration (FDA) is responsible for "protecting consumers and promoting public health." The government agency's CVM "regulates the manufacturing and distribution of food additives and drugs that will be given to animals." Site features include hot topics (mad cow disease, antimicrobial resistance, animal feed additives and ingredients, etc.); news and events; and freely accessible publications including *FDA Veterinarian* newsletter, *FDA Approved Animal Drug Products* (Green Book), guidance documents, and more (Figure 2.9).

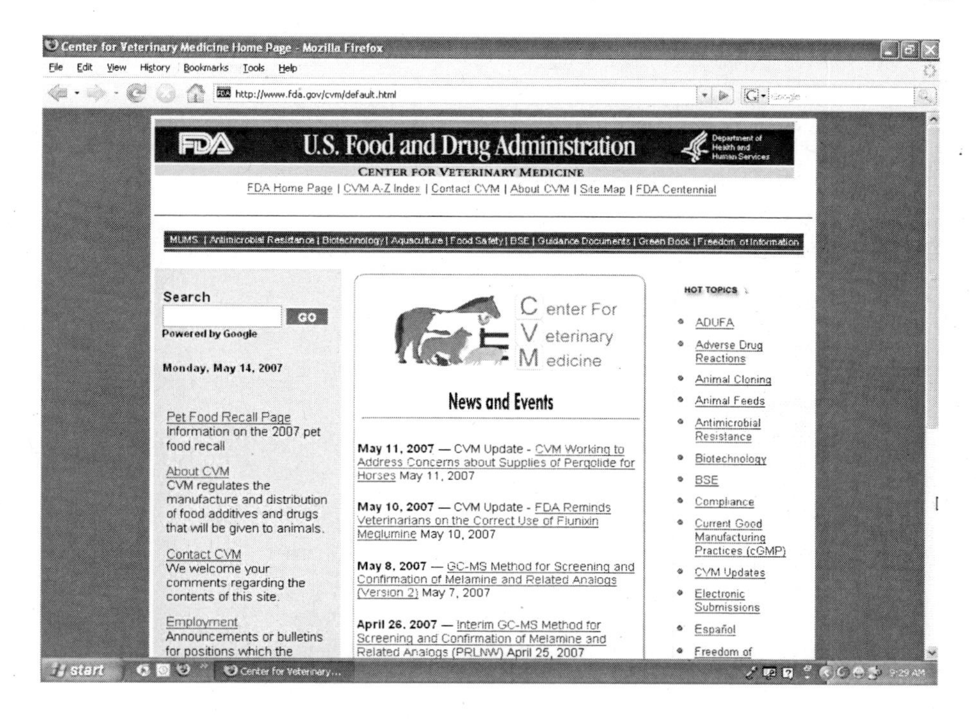

FIGURE 2.9. Center for Veterinary Medicine (CVM) <http://www.fda.gov/cvm/default.html>

Department for Environment, Food, and Rural Affairs (DEFRA) <http://www.defra.gov.uk/animalh/index.htm>

The United Kingdom's DEFRA concerns itself with economic, social, and environmental aspects of sustainable development. Site features include documentation of disease outbreaks in Europe; reportable diseases; diseases spread from animals to humans; disease monitoring, surveillance, and control; and other information.

Federation of Veterinarians of Europe (FVE) <http://www.fve.org/index.html>

FVE represents veterinary organizations in thirty-five European countries, and veterinarians in private practice, public health, and government service. The FVE site includes information and news about animal-to-human transmission of diseases, antibiotic resistance, bird influenza, foot and mouth disease, etc., and external links to veterinary organizations, agencies, and schools.

☑ The Humane Society of the United States (HSUS)
<http://www.hsus.org/>

HSUS is an organization that stresses respect and compassion for animals. Site information emphasizes humane treatment of pet animals, wildlife, farm animals, marine mammals, and research animals. Site features include pet news and events, legislation and laws, publications, e-mail alerts, finding local veterinarians and pet shelters, and much more.

☑ National Office of Animal Health (NOAH)
<http://www.noah.co.uk/>

NOAH is based in the United Kingdom and represents manufacturers in their efforts "to promote the benefits of safe, effective, quality medicines for the health and welfare of all animals." Site features include news, topics (antibiotics, vaccines, adverse reactions, etc.), pertinent legislation, links to external sites, and brief documents on a variety of subjects including vaccination of pets, antibiotic resistance, laboratory animals, and more (Figure 2.10).

Partnership for Animal Welfare (PAW)
<http://www.paw-rescue.org/>

PAW is a nonprofit organization that focuses on "the rescue and adoption of homeless domestic animals, mainly dogs and cats." This site includes information on dogs, cats, and rabbits available for adoption and/or foster care; news; pet care related to declawing, spaying, and neutering; allergies to animals; and more.

☑ Pet Health Council (PHC)
<http://www.pethealthcouncil.co.uk/>

PHC is a U.K.-based organization that focuses on the health and welfare of pet animals and works with various organizations (British Medical Association, British Small Animal Veterinary Association, British Veterinary

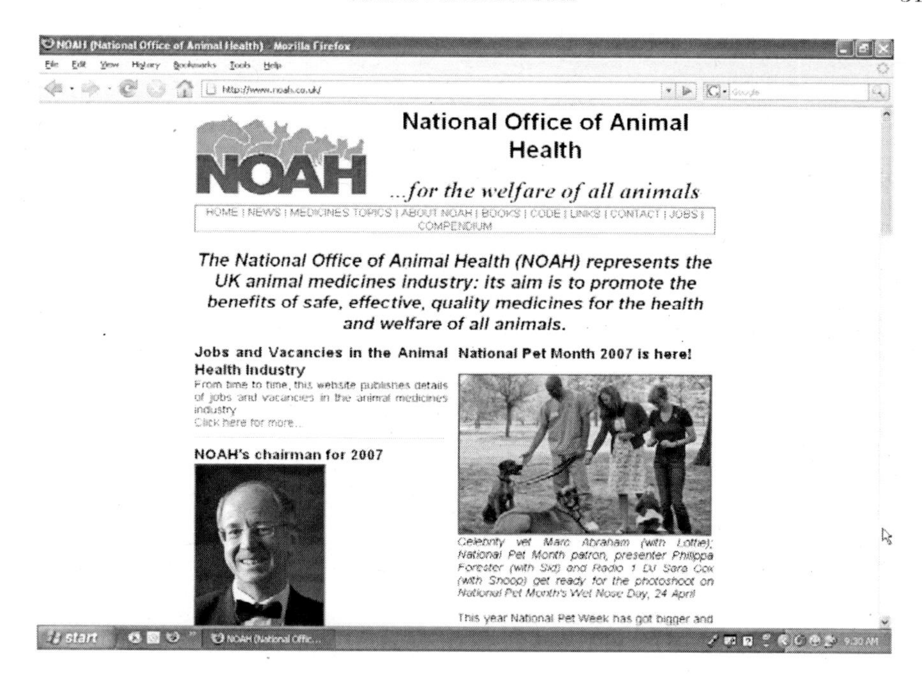

FIGURE 2.10. National Office of Animal Health (NOAH)
<http://www.noah.co.uk/>
Reprinted with permission.

Association, National Office of Animal Health, Pet Food Manufacturers' Association, Society for Companion Animal Studies, Royal College of Nursing Complementary Therapies Forum), which supply expert and authoritative content. Site features include leaflets on a variety of topics (overweight pets, toxoplasmosis, traveling overseas with pets, dental health, allergies, pet insurance, and more); health benefits of pet ownership; *Pet Health Matters* e-mail bulletin; contact information for various organizations of interest to pet owners; and an online library restricted to veterinary personnel who complete a simple registration process (Figure 2.11).

Royal College of Veterinary Surgeons (RCVS)
<http://www.rcvs.org.uk/>

RCVS "is the regulatory body for veterinary surgeons in the United Kingdom," and strives "to safeguard the health and welfare of animals

FIGURE 2.11. Pet Health Council (PHC)
<http://www.pethealthcouncil.co.uk/>
Reprinted with permission.

committed to veterinary care through the regulation of the educational, ethical, and clinical standards of the veterinary profession." Site features include news, e-mail newsletter, and a database for finding a veterinarian.

United States Animal Health Association (USAHA) <http://www.usaha.org/>

USAHA is a nonprofit organization that "works with state and federal governments, universities, veterinarians, livestock producers, national livestock and poultry organizations, research scientists, the extension service, and seven foreign countries to control livestock diseases in the United States." Site features include news alerts (mad cow disease, tuberculosis in cattle, bird influenza, etc.) and links to external sites.

PUBLICATIONS

Many useful pet health sites feature brief articles and documents of interest and value to pet owners. While more and more of the scientific literature in the form of journal articles and books have become publicly accessible on the Web, the majority of full-text content is restricted to members of a specific organization or paid subscribers.

AWIC Bulletin
<http://www.nal.usda.gov/awic/pubs/bulletin.shtml>

The United States Department of Agriculture's Animal Welfare Information Center publishes *AWIC Bulletin* that "provides current information on animal welfare to investigators, technicians, administrators, exhibitors, and the public." *Bulletin* issues are accessible through the Web site or available via e-mail. Past topics have included information about animal welfare research, alternative animal models for diabetes, whether animals feel pain, and more.

Animal Care Sheets
<http://www.petco.com/CareSheets/CareSheets.aspx?Nav=154>

PETCO, a pet product company, developed these Companion Animal Care Sheets to cover the care of dogs, cats, fish, birds, reptiles, lizards, turtles, snakes, and other small pets. A typical care sheet explains the suitability of a particular pet animal to the person adopting it, diet, bedding or housing, behavioral traits, grooming, common health issues, and much more.

☑ Current Veterinary Journals
<http://www.medvet.umontreal.ca/biblio/vetjr.html>

The University of Montreal's veterinary medical library developed this list of more than 250 current veterinary journals, instructions for authors, and impact factors. Some of the content is restricted to tables of contents or abstracts. The site also features a database of conference proceedings.

The Daily Cat
<http://www.thedailycat.com/>

The Daily Cat is an online magazine devoted to understanding cats and their behavior. Regular columns include Bringing Up Kitty (grooming, adoption, exercise, cat bedding, and toys), Your Cat's Mind (behavior problems, cat senses, training, therapy cats), Cat Culture (cat cards, interviews with cat lovers), and Cat Facts (breed profiles, cat care tips).

☑ The Green Book On-Line
<http://www.fda.gov/cvm/Green_Book/elecgbook.html>

The Center for Veterinary Medicine (CVM) at the U.S. Food and Drug Administration (FDA) has published this resource since 1989. *The Green Book* is "a list of all animal drug products approved for safety and effectiveness." Chapters cover trade names, active ingredients, patent information, voluntary withdrawals, monthly updates, and more. Content is available in hypertext and PDF formats (Figure 2.12).

Healthypet
<http://www.healthypet.com/>

The American Animal Hospital Association (AAHA) accredits animal hospitals in the United States and Canada. AAHA's Healthypet site includes a Pet Care Library with articles and frequently asked questions about dogs, cats, and other animals; pet photo page; and a database of veterinary hospitals searchable by name, state/province, ZIP code, or postal code within a certain number of miles.

☑ International Veterinary Information Services (IVIS)
<http://www.ivis.org/>

IVIS is a nonprofit organization that provides information to veterinarians, veterinary students, and others. Access to site information is free but requires a registration process. Site features include an e-mail newsletter; calendar of events (conferences and educational programs); electronic books related to birds, companion animals, cattle, horses, exotic and zoo animals, pigs, and other animals; conference proceedings; career opportunities; searchable databases; links to external sites; and more.

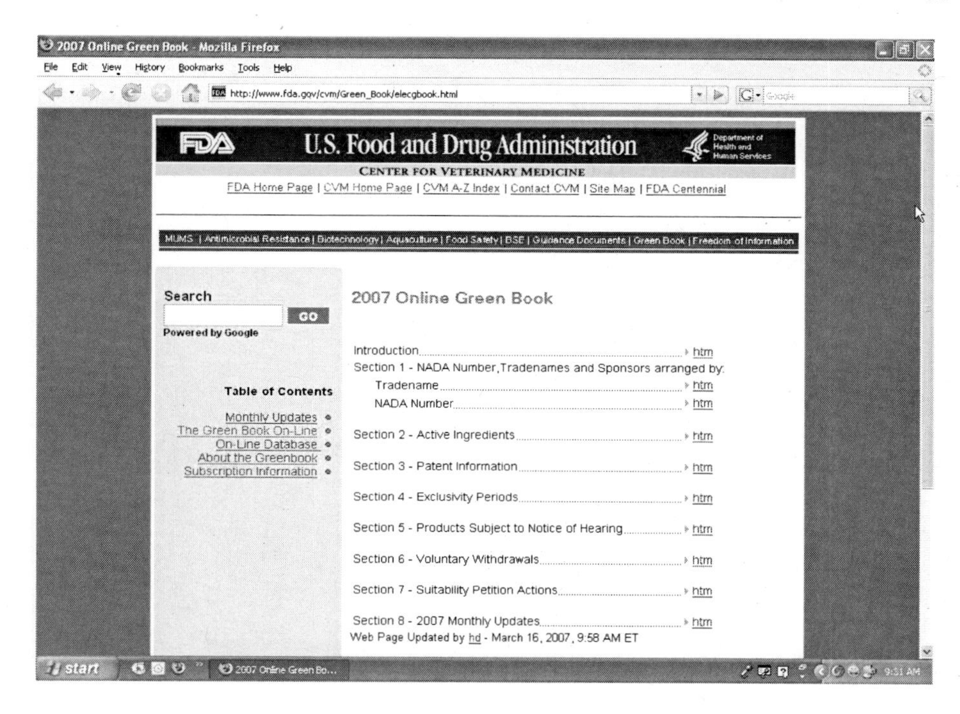

FIGURE 2.12. *The Green Book*
<http://www.fda.gov/cvm/Green_Book/elecgbook.html>

☑ The Merck Veterinary Manual <http://www.merckvetmanual.com/mvm/index.jsp>

Merck & Co., Inc. is a major pharmaceutical firm that publishes a number of manuals of use and value to medical professionals. Merck describes their veterinary manual as "the single most comprehensive electronic reference for animal care information . . . [With] over 12,000 indexed topics and over 1,200 illustrations." Coverage includes animal organ systems, behavior, emergency medicine and critical care, exotic animals, laboratory animals, nutrition, drug therapy, diseases transmitted from animals to humans, poisoning, and more. Content can be browsed through an outline or can be searched by "topic, species, specialty, disease, and keyword."

☑ VetCentric.com Encyclopedia
<http://www.vetcentric.com/reference/encyc.cfm>

VetCentric is a site that focuses on the health and behavior of dogs, cats, horses, birds, fish, ferrets, rabbits, "pocket pets" (tarantulas, hamsters, sugar gliders, guinea pigs, mice, chinchillas, etc.), reptiles, farm animals, and wildlife. Site features include an encyclopedia of common and not so common diseases, glossary of medical terms, and frequently asked pet health questions.

☑ The Veterinary Dental Encyclopedia
<http://www.dentalvet.com/Encyclopedia/P00352_SC01094.htm>

All Pets Dental is a veterinary practice based in Weston, Florida that specializes in pet dentistry. This site includes useful information about gum disease, fractured teeth, orthodontics, general dental care, and more. The site also includes a dental encyclopedia developed by veterinarian Jan Bellows, and covers topics such as dental anatomy, radiology, oral surgery, etc. Typical encyclopedia entries include photographs, radiographs, and explanatory text.

☑ Veterinary Sciences Tomorrow:
An Online Current Awareness Journal
<http://www.vetscite.org/>

Published in the Netherlands, this online-only journal publishes refereed reviews and opinions about animal health and welfare research. The typical issues included are how distemper originates and develops, embryonic stem cells, alternative medicine, and much more.

NOTE

1. Myers, John E. B. *Child Protection in America: Past, Present, and Future*. New York: Oxford University Press, 2006, p. 35.

Chapter 3

Specific Issues and Concerns for Pet Owners

ALLERGIES

Allergies to Pets
<http://www.hsus.org/pets/pet_care/allergies_to_pets/>

The Humane Society of the United States (HSUS) focuses on the humane and compassionate treatment of pets, wildlife, farm animals, laboratory animals, and marine mammals. The information about pet allergies covers ways to alleviate allergic reactions to dogs and cats.

Animal Allergy
<http://www.animalhealthchannel.com/animalallergy/>

Animalhealthchannel "provides comprehensive, trustworthy information about conditions and diseases that affect animals, as well as pet care and training information," and its content is "developed and monitored by board-certified veterinarians." The pet allergy content covers an overview of atopy, allergens, causes, signs and symptoms, diagnosis, treatment, and prevention. Helpful tips on allergy prevention (based on recommendations made by the Asthma and Allergy Foundation of America) include using HEPA filters, brushing and washing pets regularly, covering air vents with cheesecloth, using bare floors rather than carpets, and keeping pet animals out of human sleeping areas.

Internet Guide to Pet Health

doi:10.1300/5675_03

Animals and Insects Center—Allergy and Asthma
<http://allergy.health.ivillage.com/animalsinsects/>

Animals and Insects Center includes information about allergies to cats, dogs, birds, and other small animals, insect stings, and more. It is written by physicians specializing in allergy and hypersensitivity. This site's information covers causes, symptoms, diagnosis, treatment, and research into pet allergies.

Pet Allergies—Healthology
<http://www.healthology.com/focus_article.asp?f=allergy&c=petallergies>

Healthology is a site that distributes health information generated by physicians. The pet allergies content covers diagnosis and treatment for persons sensitive to dogs, cats, horses, rodents, cockroaches, and dust mites.

Pet Allergies—Intelihealth
<http://www.intelihealth.com/IH/ihtIH/WSIHW000/7945/8213.html>

Intelihealth is a site that provides consumer health information developed by experts associated with Harvard Medical School. The site's information about pet allergies explains cat and dog dander, and its role in causing human allergic symptoms.

Pet Allergies—NetDoctor
<http://www.netdoctor.co.uk/health_advice/facts/allergypets.htm>

NetDoctor is a U.K.-based resource that offers information and advice about a variety of conditions and diseases. The section about pet allergies explains why humans are sensitive to pet allergens, specific allergic reactions (such as sneezing, runny nose, wheezing, watery eyes, itching), household cleaning methods, and drug treatments.

Pet Allergy
<http://www.allergysa.org/pet.htm>

The Allergy Society of South Africa's site explains how humans develop allergic reactions to dogs, cats, small pets, and farm animals through animal saliva, skin, or urine, and how antihistamines and corticosteroids can be used to treat symptoms.

DEATH, DYING, AND EUTHANASIA

Since most pet animals have far shorter life spans than average humans do, pet owners may face the dilemma of caring for ill and dying pets several times over the years. Chronic illnesses may require pet owners to consider euthanasia for their beloved companions.

Association for Pet Loss and Bereavement (APLB)
<http://www.aplb.org/>

APLB is a Brooklyn, New York-based nonprofit organization that is comprised of persons who have experienced pet loss. The APLB site includes a bibliography of publications available from Amazon.com, chat room, electronic condolence cards, newsletters, tributes to pets who have passed away, listing of pet cemeteries and cremation facilities by state, and topics such as euthanasia.

Care for Animals
<http://www.avma.org/careforanimals/animatedjourneys/goodbyefriend/goodbye.asp>

The American Veterinary Medical Association (AVMA) provides guidance for deciding to have a sick or injured pet euthanized by a veterinarian.

Euthanasia—Introduction
<http://www.merckvetmanual.com/mvm/index.jsp?cfile=htm/bc/150500.htm&word=euthanasia>

Drug manufacturer Merck & Co., Inc.'s veterinary manual includes more than 12,000 topics including euthanasia. This section describes the use of inhaled anesthetic agents, injections, and other humane methods.

Euthanasia . . . What to Expect
<http://www.thepetcenter.com/imtop/euthanasia.htm>

PetFoodDirect.com owns and operates this site with content developed by practicing veterinarians. This site functions as a virtual animal hospital with content and images that explain physical examination, radiology, surgery, and drug therapy for diseases and conditions common to dogs and cats. The information about euthanasia discusses making an appointment, what to expect, in-home euthanasia options, and handling grief.

An Introduction to Pets in Wills and Pet Euthanasia
<http://www.animallaw.info/articles/arusdanielssuzette2004.htm>

Michigan State University's College of Law site includes an Animal Legal and Historical Web Center. The information covers pet trusts, will-stipulated pet euthanasia, and estate planning.

Pet Euthanasia
<http://www.cvm.uiuc.edu/petcolumns/showarticle.cfm?id=493>

University of Illinois Urbana-Champaign's College of Veterinary Medicine site includes useful information about the difficult decision to put a pet to sleep and how to handle grief.

Pet Euthanasia
<http://www.thefuneraldirectory.com/eu_expect.html>

The Funeral Directory site has a section devoted to pets that covers pet euthanasia, pet memorials, discussion forum, and dealing with grief.

Pet Loss
<http://www.petvets.com/petloss/>

Harmony Animal Hospital's World of Pets site includes pet postcards; photo contest; free newsletter; pet care information for owners of dogs, cats, birds, rabbits, or reptiles; and a substantial section about pet loss including saying goodbye, grieving, and sharing stories.

Pet Loss Hotline—Cornell University
<http://www.vet.cornell.edu/Org/PetLoss/>

Cornell University's Pet Loss Support Hotline is staffed by volunteer veterinary students. The site includes information about the hotline service, pet euthanasia, other pet loss hotlines staffed by veterinary students, stages of grieving for deceased pets, and links to other sites related to pet loss.

Pet Loss Hotline—Washington State University
<http://www.vetmed.wsu.edu/PLHL/>

Washington State University's College of Veterinary Medicine site includes pet health topics (bunny care, adopting a pet, determining the sex of a kitten, etc.), anatomy (text and images related to dog or cat organ systems), and definitions and abbreviations of common veterinary terms. The pet loss information stresses that it is all right "to love and miss your pet," and includes a hotline service staffed by veterinary students.

Pet Loss Sympathy Cards
<http://www.geocities.com/sandytracker/plcards.html>

This site links to a variety of pet loss sympathy cards including ones that are specifically designed for cats, dogs, ferrets, rabbits, horses, rats or mice, and birds. These cards can be personalized and sent to specified e-mail addresses.

When to Consider Euthanasia in Dogs
<http://www.petplace.com/articles/artShow.asp?artID=990>

Veterinarian Jon Rappaport's site features separate pet centers for cats, dogs, birds, horses, reptiles, fish, and small mammals; drug library; medical center; veterinarian locator; and information about pet insurance. The information about dog euthanasia covers quality of life, effects of medication, and costs of veterinary care.

DENTAL CARE

Dental health is an important aspect of pet care. Just like their human guardians, dental health starts with regular brushing, consumption of healthy and crunchy foods, avoidance of foods that cause tooth decay, fractures, and loss; regular checkups, and treatment as needed. See also Teeth and Gums in the cat or dog sections for additional links.

All Pets Dental
<http://www.dentalvet.com/>

All Pets Dental is a veterinary practice based in Weston, Florida that specializes in pet dentistry. The practice's site includes useful information about gum disease, fractured teeth, orthodontics, dentistry for cats, general dental care, and more.

Common Problems
<http://www.dogbeachdentistry.com/problems.html>

Dog Beach Dentistry is a site developed by veterinarian Brook Niemiec. Site information includes brief explanations and images of various dental problems (dead, fractured, decayed teeth) and jaw diseases (tumors, fractures, fistulae, gum infections) mostly in dogs.

Dental Disease in Dogs and Cats
<http://www.vetstop.com.au/info/Dental.asp>

VetStop sells a variety of products for pet owners including flea treatments, toys, pet care supplies, treats, etc. The information about dental disease in dogs and cats covers prevention (meaty bones, chew toys, tooth brushing, and dry foods).

Dental Health
<http://www.intute.ac.uk/healthandlifesciences/cgi-bin/browse.pl?gateway=vetgate&id=57131>

Intute, formerly known as VetGate, is a veterinary gateway that links to useful external sites, including information about periodontal disease, root canals for fractured teeth, orthodontics, and preventive dental care for dogs, cats, and horses.

Dental Home Care
<http://www.marvistavet.com/html/dental_home_care.html>

The Mar Vista Animal Medical Center in Los Angeles, California offers a wide range of pet health information on their site. The information about dental care includes toothpaste, toothbrushes, dental sprays, and dental treats that can be used with pets.

Periodontal Disease
<http://www.merckvetmanual.com/mvm/index.jsp?cfile=htm/bc/20408.htm&word=periodontal>

Drug manufacturer Merck & Co., Inc.'s veterinary manual includes more than 12,000 topics including periodontal disease, which is an infectious process of the gums and bone surrounding the teeth. This section explains the causes, treatment, and prevention of this disease, and also explains the difference between periodontal disease and gingivitis.

DIET AND NUTRITION

The health of companion animals depends greatly on diet and nutrition. The Association of American Feed Control Officials (AAFCO) established nutrient profiles <http://www.aafco.org/> for dog food and cat food, which require specific amounts of various nutrients in order to be considered complete and balanced. Pay close attention to food labels, ingredients, portion sizes, and your pet's nutritional requirements, based on breed, age, and physical activity.

Diets and Nutrition for Cats, Dogs, and Other Pets
<http://animalpetdoctor.homestead.com/Nutrition.html>

Roger Ross is a veterinarian who practices animal medicine in Seneca, South Carolina. This site includes easy to understand and useful information about the treatment of various diseases and conditions commonly seen in cats and dogs. The site includes information about diet and nutrition and covers supplements, vitamins, and minerals; alternative diets; "natural" pet foods; prescription diets; and more.

Health and Nutrition
<http://www.canadasguidetodogs.com/health.htm>

Canada's Guide to Dogs describes its Web site as "a completely independent project," not affiliated with any company or professional association.

Site sections include dog breeds and breeders, activities and clubs, working dogs, training and handling, health and nutrition, dog walking and pet sitting, grooming and groomers, products and retailers, dog rescue organizations and shelters, dog judges, and dog shows and events. The health and nutrition section features an alphabetic list of links related to dog health.

Hill's Pet Nutrition
<http://www.hillspet.com/>

Hill's Pet Nutrition, Inc. manufactures food for dogs and cats. Their pet nutrition site focuses on product information for pet owners, persons working in veterinary practices, and pet food retailers. Site features include information about their pet foods, puppy and kitten care tips, an interactive game about pet myths, and database of veterinarians searchable by ZIP code, city, state, etc.

Nutrition Center
<http://www.familypetservices.com/health_nutcntr_faq.asp>

Family Pet Services is a site that describes itself as a "24-hour online pet care reference." Nutrition information covers ingredients in cat food or dog food, feeding times for dogs, and frequently asked questions.

Pet Health—Nutrition
<http://www.healthcyclopedia.com/animal/pets/nutrition.html>

Healthcyclopedia is a site that provides brief information about a variety of health matters, including pet health. Subtopics include nutrition, emergency services, and publications. Each subtopic, animal type, or subject is further subdivided into subtopics. For example, the section about nutrition is subdivided into alternative; commercial pet food; carbohydrates, sugars, fiber; fats, oils, fatty acids; toxic foods; vitamins and minerals; and more.

Pet Nutrition
<http://www.thepetcenter.com/imtop/nutritioncomments.html>

PetFoodDirect.com owns and operates this site with content developed by practicing veterinarians. This site functions as a virtual animal hospital with content and images that explain physical examination, radiology, surgery, and drug therapy for diseases and conditions common to dogs and cats. The information about nutrition covers the inadequacies of grain-based pet food, nutritional composition of bones, dangers of pets eating bones, determining whether a cat is overweight, and more.

Pet Owner Home
<http://www.sciencediet.com/>

Science Diet is a brand of pet food manufactured by Hill's Pet Nutrition, Inc. This information about pet ownership covers nutrition, care of puppies and kittens, finding a veterinarian, superstar pets of the month, pet myths, and special diet needs of dogs and cats.

☑ **PetDIETS.com**
<http://www.petdiets.com/>

This site dispenses nutritional advice from persons who specialize in veterinary nutrition. Site features include "Pet Nutrition 101," which covers nutritional stages (growth, maintenance, old age); daily energy requirements for adult cats and dogs based on weight; a food calculator for determining the amount of canned or dry food required by an individual adult cat or dog; facts and myths about pet food ingredients; advantages and disadvantages of homemade pet foods; and more.

DISABLED PETS

Pets may be born with various birth defects or suffer debilitating injuries later in life that require special care. Some organizations focus on rescuing and providing foster and adoptive homes for such pets, and others provide assistance and/or equipment to persons who have adopted pets that have disabilities.

Deaf Dog Friendly Rescues and Shelters
<http://www.deafdogs.org/rescue/>

Deaf Dog Education Action Fund is a nonprofit organization that focuses on promoting spaying/neutering, eliminating euthanasia, and arranging rescues for deaf dogs. Site information includes training tips, recommended reading, and listings of deaf dogs for foster care or adoption.

☑ Handicapped Pets <http://www.handicappedpets.com/>

Mark Robinson developed this site to honor his pet keeshound, Mercedes, who was euthanized for epilepsy a number of years ago (Figure 3.1). The organization's stated mission is to "support the caretakers of Elderly, Disabled, and Handicapped pets." Site features include product information (wheelchairs, bedding, booties, pet diapers, etc.), services (counseling, animal chiropractors, and holistic health practitioners, etc.), and support (discussion boards, galleries of images, memorials, pet rehabilitation facilities).

FIGURE 3.1. Handicapped Pets
<http://www.handicappedpets.com/>
Reprinted with permission.

Living with a Disabled Pet
<http://www.bestfriends.org/theanimals/pdfs/allpets/disabledpet.pdf>

The Best Friends Animal Society is located in Kanab, Utah. Veterinarian Virginia Clemans developed this two-page color document about caring for pets with disabilities such as missing limbs, arthritis, blindness, and deafness.

Living with a Hearing Impaired Dog
<http://www.animalhealthcare.ca/contents/content_careguides2.asp?get=faq&id=128&category=44>

Canadian Veterinary Medical Association provides information about the challenges of owning a dog that has a hearing impairment. Tips include using positive reinforcement, smell, and hand signals to train and reward, and safety reminders.

Pets with Disabilities—Healthypet.com
<http://www.healthypet.com/library_view.aspx?ID=10&sid=1>

American Animal Hospital Association's Healthypet.com site includes compelling information about pets with disabilities such as chronic diseases (arthritis, heart disease, and kidney failure), blindness; deafness; or missing, deformed, or paralyzed limbs.

Pets with Disabilities
<http://www.petswithdisabilities.org/>

Petswithdisabilities.org was founded by Joyce and Michael Dickerson of Prince Frederick, Maryland in an effort to "raise awareness about disabled pets." Site features include inspirational stories and pictures of various disabled pets, disabled pets for adoption, foster programs for disabled cats or dogs, and more.

Wheelchairs for Pets
<http://www.eddieswheels.com/>

An engineer named Eddie fashioned a set of wheels for Buddha, a Doberman pinscher who had lost the use of his rear legs. This site includes a picture gallery of dogs kept active through the use of this equipment, and some stories about dogs that have been helped (Figure 3.2).

EMERGENCY CARE AND FIRST AID

Pets need effective and immediate care for injuries and health emergencies, just as humans do. Some situations can be handled at home with basic first aid supplies but others will require expert veterinary care.

Eye Diseases <http://www.animaleyecare.net/diseases/diseases.htm>

Animal Eye Care LLC is located in Bellingham, Washington and provides eye care services to dogs, cats, birds, gorillas, horses, and other

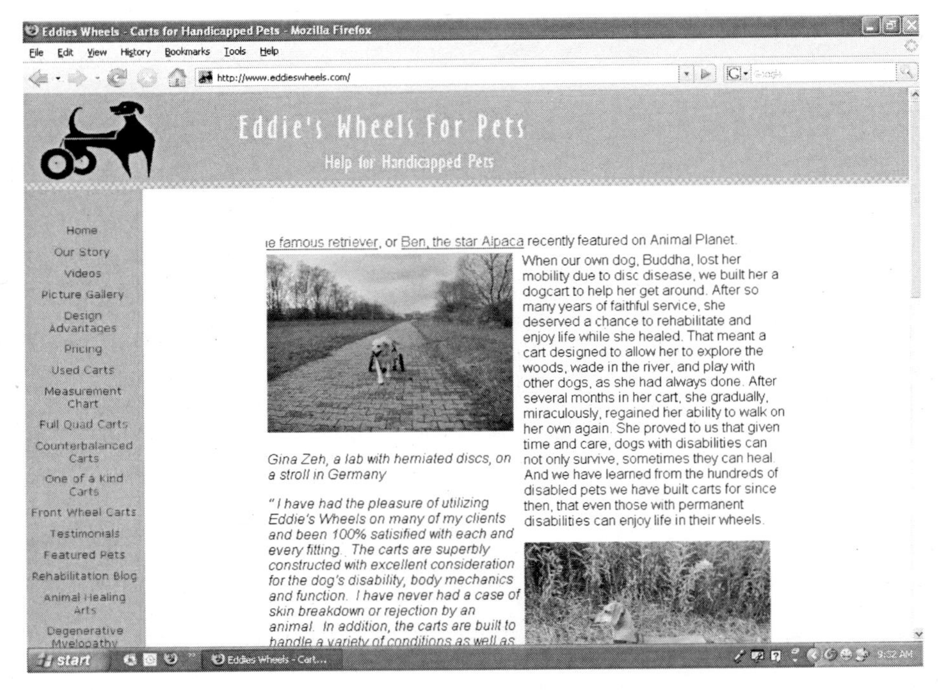

FIGURE 3.2. Eddie's Wheels—Carts for Handicapped Pets
<http://www.eddieswheels.com/>
Reprinted with permission.

animals. Site features include an alphabetic listing of eye diseases from cataracts to glaucoma to uveitis (eye inflammation).

Emergencies and Injuries
<http://www.animalhealthchannel.com/emergency/>

Animalhealthchannel provides "comprehensive, trustworthy information about conditions and diseases that affect animals, as well as pet care and training information," and its content is "developed and monitored by board-certified veterinarians." The information about pet emergencies and injuries covers bites, burns, breathing difficulties, diarrhea, fractures, heatstroke, poisoning, seizures, shock, vomiting, and necessary emergency supplies.

Fight Wound Infections
<http://www.sniksnak.com/cathealth/fight-wounds.html>

Pawprints and Purrs, Inc. is a Louisiana-based nonprofit organization that concerns itself with the welfare and well-being of cats and dogs. The information about wound infections in cats covers clinical signs, diagnosis, treatment, risks to humans, and more.

First Aid Kit for Cats
<http://www.petplace.com/cats/first-aid-kit-for-cats/page1.aspx>

Veterinarian Jon Rappaport's site information about first aid for cats covers preparing a kit of first aid supplies, cardiopulmonary resuscitation (CPR) in cats, determining whether there is a medical emergency, and much more.

First Aid Kit for Dogs
<http://www.petplace.com/dogs/first-aid-kit-for-dogs/page1.aspx>

Veterinarian Jon Rappaport's site information about first aid for dogs covers preparing a kit of first aid supplies, performing cardiopulmonary resuscitation (CPR) on dogs, using the Heimlich maneuver to save a choking dog, determining whether there is a medical emergency, and much more.

Pet Emergencies
<http://www.bringyourpet.com/emergencies.htm>

BringYourPet.com is a site that focuses on pet-friendly lodging available in the United States, Canada, and other parts of the world. The information about preparing for pet emergencies while away from home features a list of first aid supplies, emergency contact numbers, and deciding whether to consult a veterinarian.

Pet First Aid and Emergency Care
<http://www.thepetcenter.com/gen/faid.html>

PetFoodDirect.com owns and operates this site with content developed by practicing veterinarians. This site functions as a virtual animal hospital with content and images that explain physical examination, radiology, surgery, and drug therapy for diseases and conditions common to dogs and cats. The brief information about first aid covers cuts, scrapes, burns, frostbite, fractures, hot spots (moist eczema), bites, poisoning, and other emergencies.

Pet Resources-Emergency Services
<http://www.petsonthego.com/resourcesemerg.html>

Pets on the Go! is a site that focuses on products and services for people who like to travel with their pets. Site features include information about pet-friendly accommodations (motels, hotels, resorts, and rental properties), pet-friendly transportation (car rentals, air travel, and cruise companies), pet movers, pet sitters, preparing for and dealing with emergencies, and more. The section about emergency services gives tips for natural disasters and other emergencies including preparing a first aid kit, what to do if a pet needs to be left behind, and a list of emergency contacts.

Protecting Pets in Everyday Emergencies
<http://www.hsus.org/hsus_field/hsus_disaster_center/disaster_preparedness_for_pets/protecting_pets_in_everyday_emergencies.html>

The Humane Society of the United States (HSUS) focuses on the humane and compassionate treatment of pets, wildlife, farm animals, laboratory animals, and marine mammals. The information about everyday emergencies covers planning that is necessary for unexpected situations.

EXERCISE AND TRAINING FOR CATS AND DOGS

Regular exercise is important for many pet animals, large and small. Training is essential for dogs while some cats may be amenable to such activities. Please refer to the sections related to behavior problems of cats or dogs elsewhere in this book.

Dog Exercise
<http://www.petguide.org/dogexercise.htm>

PetGuide.org focuses on food, water, treats, baths, bedding, medical care, exercise, and toys for pet dogs and cats. The information about dog exercise offers commonsense and practical advice for keeping your dog happy and healthy.

DogPlay—Activities for Dogs
<http://www.dogplay.com/index.html>

DogPlay is a site that was developed by Diane Blackman to sell playful dog-related products and promote dog play. The site links to various play activities ranging from agility, flying disc, and fly ball to training camp, sledding, hiking, and much more.

Dog Training, Puppy Training, Cat and Kitten Training and Behavior
<http://www.perfectpaws.com/>

Perfect Paws includes detailed information about puppy, dog, kitten, and cat training. The sections related to training dogs feature the use of leashes, crates, and play activity to train and socialize. The parts about training cats feature the use of litter boxes, scratching posts, and toys to curb inappropriate elimination, scratching, and excessive vocalization.

☑ Dog Whisperer with Cesar Millan <http://channel.nationalgeographic.com/channel/dogwhisperer/index.html>

The National Geographic Channel features the immensely popular dog behavior specialist Cesar Millan who emphasizes exercise, discipline, and affection. The Dog Whisperer program's Web site includes a blog, images, video clips, episode transcripts, and a sign-up for weekly behavior tips to be delivered to a mobile telephone.

Importance of Exercise <http://www.sfspca.org/behavior/cat_library/exercise.pdf>

The San Francisco Society for The Prevention of Cruelty to Animals developed this three-page document that describes the importance of exercise for cats. The information presented here stresses the importance of regular physical activity for indoor cats, health risks caused by obesity (diabetes mellitus, arthritis, cardiovascular diseases), and the physical and behavioral benefits of exercise.

Leash Training Your Cat <http://www.hdw-inc.com/leashtraining.htm>

A breeder of Bengal and Savannah cats offers this advice for leash training cats through the use of patience, a harness-like jacket and leash, lures, and treats.

Pet Training and Care <http://bloomington.in.gov/animalshelter/training.php>

The City of Bloomington, Indiana's Department of Animal Care and Control offers helpful tips about keeping cats indoors, spaying and neutering, separation anxiety in dogs, litterbox issues for cats, effective use of crates to train dogs, and much more.

Web Site for Cats <http://www.cattv.com/catwebsite.htm>

This unconventional site can be used to play motion games for cats to follow on the computer screen. Cartoon animation choices include a bluebird, butterfly, bee, bat, groundhog, and bug.

GROOMING

Good hygiene and grooming are important aspects of healthy lifestyles for pet animals. Grooming practices are often breed- and species-specific. Consult the following sites for the most useful information to keep your cat or dog fresh and healthy.

Choosing a Groomer
<http://www.hsus.org/pets/pet_care/choosing_a_groomer.html>

The Humane Society of the United States (HSUS) focuses on the humane and compassionate treatment of pets, wildlife, farm animals, laboratory animals, and marine mammals. The information about pet grooming covers choosing a groomer, evaluating a grooming facility, easing a pet's fears, and preparing for the first grooming visit.

Dog and Cat Grooming
<http://www.thepetcenter.com/groom/gr.html>

PetFoodDirect.com owns and operates this site with content developed by practicing veterinarians. This site functions as a virtual animal hospital with content and images that explain physical examination, radiology, surgery, and drug therapy for diseases and conditions common to dogs and cats. Veterinarian T. J. Dunn, Jr.'s site information covers touch, appearance, and smell of the coat; condition of ears, eyes, and nails; and avoiding cuts and abrasions.

Grooming Your Cat
<http://www.petplace.com/cats/grooming-your-cat/page1.aspx>

Veterinarian Jon Rappaport's site information about grooming and skin care for cats includes grooming kittens and senior cats, grooming supplies, removing fur mats, cleaning a cat's ears, trimming toenails, cat bathing, and more.

Grooming Your Pet
<http://www.familypetservices.com/health_groom.asp>

Family Pet Services is a site that describes itself as a "24-hour online pet care reference." The grooming information covers equipment to bathe pets and how to care for coats, nails, and teeth.

HEALTH BENEFITS OF PET OWNERSHIP

Dog and Cat Owners Gain Health Benefits
<http://www.bestfriendspetcare.com/bf_feature_14.cfm>

BestFriendsPetCare.com is a site that focuses on boarding, training, and grooming. Its Pet Library covers topics ranging from behavior/training to pet travel to lost/missing pets. The information about the health benefits of pet ownership mentions studies that have found lower blood pressure and cholesterol levels, fewer visits to physicians, higher survival rates after serious illness, reduced levels of loneliness and depression.

Dog Walking—The Health Benefits
<http://www.betterhealth.vic.gov.au/bhcv2/bhcarticles.nsf/pages/Dog_walking_the_health_benefits?open>

Better Health Channel is a site established by the Victorian Government (Australia) as a means to disseminate information about healthy living. The information about walking a dog refers to research about health, community, and the exercise benefits of doing so.

Health Benefits of Pets for People
<http://www.merckvetmanual.com/mvm/index.jsp?cfile=htm/bc/140302.htm>

Drug manufacturer Merck & Co., Inc.'s veterinary manual is "the single most comprehensive electronic reference for animal care information . . . [With] over 12,000 indexed topics and over 1200 illustrations." The section about the health benefits of pets covers relaxation, entertainment, companionship, and how pets can foster social interactions.

Healthy Pets Healthy People—Health Benefits of Pets
<http://www.cdc.gov/healthypets/health_benefits.htm>

The U.S. Centers for Disease Control and Prevention (CDC) is concerned with the monitoring and prevention of disease. Site information about diseases transmitted from animals to humans (zoonosis) includes a section browsable by animal (birds, cats, dogs, farm animals, fish, horses, reptiles, wildlife) or disease (mad cow disease, brucellosis, campylobacteriosis, cat scratch fever, *E.coli,* giardiasis, rabies, roundworm, toxoplasmosis, West

Nile virus, etc.). The information about the health benefits of pet ownership refers to research studies that show that loneliness and blood pressure, triglyceride, and cholesterol levels can be reduced.

Pet Therapy
<http://www.holisticonline.com/Pets/default.htm> browsable

HolisticOnline is an alternative medicine site developed by psychiatrist Shila Mathew and her husband Dr. Jacob Mathew. The section related to pet therapy covers the health benefits of interactions with pets including therapeutic horseback riding, better heart health through dog ownership, and benefits to children who interact with pet animals.

Physical and Medical Health Benefits
<http://www.peteducation.com/article.cfm?cls=0&cat=1278&articleid=638>

Veterinarians Foster and Smith developed this excellent resource for pet owners of dogs, cats, fish, birds, ferrets, reptiles, and other small pets. The information about the health benefits of pet ownership or interaction with pets covers longevity after heart attacks; reduced stress; lower cholesterol and triglyceride levels; lower blood pressure readings; prediction of seizure activity; diagnosis of cancer; increased physical activity; assisting persons with Parkinson's disease; and alerting persons with low blood sugar levels.

Researchers Split on Health Benefits of Pet Ownership
<http://www.vetcentric.com/magazine/magazineArticle.cfm?ARTICLEID=889>

VetCentric is a site that focuses on the health and behavior of dogs, cats, horses, birds, fish, ferrets, rabbits, "pocket pets," reptiles, farm animals, and wildlife. This article explains why researchers have differing opinions about the relative health value of owning pets, and discusses additional research that is needed.

PET INSURANCE

Pet insurance is health insurance for pet animals and covers the gamut from preventive medications for heartworm, fleas, and ticks to routine checkups, surgery, hospitalization, and treatments. According to market researchers,[1] North American pet insurance revenues are expected to reach more than $600 million by 2010, although at the present time, the percentage of North American pet insurance holders is in the single digits compared to Great Britain and Sweden. Some employers have started to include pet health coverage as a benefits option.

Embrace Pet Insurance Blog
<http://embracepetinsurance.typepad.com/blog/>

Actuary Laura Bennett runs this Web log that enthusiastically recommends pet insurance for pet owners, and explains the ins and outs of different plans, and links to insurance carriers who offer such coverage.

Pet Care Information Tips—Responsible Pet Ownership: Pet Health Insurance
<http://www.animed.org/responsible_pet_insurance.htm>

Ani-Med is a site sponsored by the American Society for the Prevention of Cruelty to Animals (ASPCA) that has two different sections: one with information about dogs and cats, and another for birds, fish, rabbits, ferrets, guinea pigs, hamsters, mice, gerbils, and rats. The information about insuring the health of pets covers typical coverage options.

Pethealthcare
<http://www.pethealthcare.co.uk/>

Pethealthcare is a site developed by an insurance company. Site information includes dog care and breeds, cat care and breeds, choosing first pets, latest news, veterinarians and rescue centers in the United Kingdom, pet health or behavior questions answered by experts, and a personal file area for keeping vaccination and other health records for pets, accessible by personal password. The information about pet insurance explains the types of veterinary charges that policies typically cover, alternative therapies, and obtaining price quotes.

POISONS, DANGEROUS SITUATIONS, AND OTHER HAZARDS

Animal Poison Control Center—A Poison Safe Home
<http://www.aspca.org/site/PageServer?pagename=pro_apcc_poisonsafe>

The American Society for the Prevention of Cruelty to Animals (ASPCA) offers tips for poison-proofing a residence such as keeping pets (especially dogs) away from foods that contain avocado, chocolate, onions, raisins, grapes, garlic; backyard hazards such as cocoa mulch, Citronella candles, and plant bulbs; over-the-counter medications (various painkillers, cold medications, vitamins); and household items (mothballs, dryer sheets).

Chocolate Toxicity
<http://www.marvistavet.com/html/chocolate_toxicity.html>

The Mar Vista Animal Medical Center in Los Angeles, California offers a wide range of pet health information on their site, including explanation of chocolate toxicity in dogs.

Cocoa Bean Mulch
<http://www.peteducation.com/article.cfm?articleid=3410>

Veterinarians Foster and Smith's site explains that the cocoa bean mulch that is used to landscape yards can be toxic to dogs.

Healthy Dogs—Poisons and Prevention
<http://www.healthy-dogs.net/poisonsprevention.htm>

Healthy Dogs focuses on dog health, care, and behavior problems. The information on preventing poisoning in dogs covers poisoning from household products, salmonella-infected foods, and pesticides. The list of poisoning symptoms (abdominal pain, salivating, vomiting, lethargy, diarrhea, burns around the mouth, skin reddening, staggered gait, twitching, depression, convulsions, or coma) is useful.

Household Hazards and Poisons
<http://www.safety.com/articles/household-hazards-and-poisons.html>

Safety.com refers to itself as the "official safety site of the Internet." The information about pets includes choking (particularly washcloths), toxins (cleaning products, particularly ones with lye), and other dangers (open dryers, washers, microwave ovens, etc.).

Household Poisons
<http://www.bellaonline.com/articles/art7240.asp>

BellaOnline is a site that focuses on topics of potential interest to women, including a help sheet of household products that are poisonous to dogs and what to do if a dog ingests any of these poisons. Dangers include antifreeze, arsenic in rat poisons, bleach, nail polish, nail polish remover, shampoo, shoe polish, turpentine, and various cleaners.

Household Poisons and Your Pet
<http://www.brentwoodpetclinic.com/householdpoisons.html>

Brentwood Pet Clinic is located in West Los Angeles, California. The clinic's site lists pet dangers including antifreeze (cats and dogs), secondhand smoke (all pets), nicotine in ashtray residues (birds), chocolate (dogs), walnuts (dogs), baby food if it contains onions (cats), specific types of houseplants (cats and rabbits), lead (birds and rabbits), Teflon (birds), rubber bands and erasers (ferrets), and ammonia and bleach (fish).

Household Poisons—Cat Guide
<http://animal.discovery.com/guides/cats/safety/poisonsintro.html>

Animal Planet features a cat guide that lists household products that are poisonous to cats including cleaning products, pain relievers, deodorants, detergents, chocolate, shampoo, and other common chemicals.

Indiana Plants Poisonous to Livestock and Pets
<http://vet.purdue.edu/depts/addl/toxic/cover1.htm>

Purdue University's Cooperative Extension Service's database of poisonous plants can be browsed alphabetically by plant name, toxicity rating (extremely toxic to minimally toxic), animal species affected (all animals, pets, livestock, and horses), botanical type (grasses, houseplants, other non-woody plants, woody bushes/trees), and line drawing of specific plants. The site's disclaimer states, "This material is for informational purposes only, not as a guide to home treatment."

☑ The Pet Friendly House—Living in Harmony with Dogs, Cats, Birds, and Fish
<http://www.petfriendlyhouse.com/>

The Pet Friendly House site includes sections on children and pets (tips for meeting a dog, preventing bites, introducing the baby to the dog), dogs (fear of thunder, dog breath, separation anxiety), plants and gardens (non-toxic plants, pesticides, dogs given as holiday gifts), and cats (whether to declaw, fearfulness, why cats bite). Other features include video clips, discussion forums, and progress on their plans to build a totally pet-friendly solar-powered house.

Pet Poisoning
<http://www.dr-dan.com/newpage16.htm>

Dr. Dan is a Marietta, Ohio-based veterinarian whose site provides useful information to pet owners. The information about pet poisoning covers poisons absorbed into the skin, ingested, or inhaled, lists many common plants that can prove poisonous, and stresses the importance of contacting the National Animal Poison Control Center at 1-800-548-2423 or 1-900-680-0000.

Pets and Plants
<http://www.mitoxic.org/pcc/petsplants/>

The Poison Control Center at Children's Hospital of Michigan provides information about poison-proofing a home to safeguard pets; hazardous foods, plants, and chemicals that can affect pets; how to react and respond to pet poisonings; and an alphabetically arranged list of plants poisonous to cats and dogs.

Plants Poisonous to Dogs
<http://www.uexplore.com/health/poisonplants.htm>

DogPack.com is a pet food distributor located in Dallas/Fort Worth, Texas. Their site information about poisonous plants was derived from information developed by the ASPCA Animal Poison Control Center. The listing is organized by plant type (bulbs, ferns, flowering plants, perennials, houseplants, lilies, shrubs, succulents, trees, vines, and miscellaneous). The site also links to external information on this subject.

Plants to Avoid—Cat Guide
<http://animal.discovery.com/guides/cats/safety/plantsavoidintro.html>

Animal Planet features a cat guide that lists plants that are harmful to cats including various flowering bulbs (amaryllis, iris, tulip, daffodil), fruits and berries (peach, cherry, blackberry), flowering bushes (oleander, azalea, hydrangea), and many others.

Poisonous Plants Informational Database
<http://www.ansci.cornell.edu/plants/index.html>

Cornell University's College of Veterinary Medicine developed this database about plants that can be harmful to animals. Information is organized alphabetically by botanical name, common plant name, toxic agent, and animal species (humans, livestock, cats, dogs, iguanas, rabbits), and also links to external information about non-plant toxic substances. Despite the wealth of information included on Cornell's site, site navigation makes it difficult to access this information readily.

Poisons Affecting Pets (Dogs and Cats)
<http://www.thepetcenter.com/wai/poi.html>

PetFoodDirect.com owns and operates this site with content developed by practicing veterinarians. This site functions as a virtual animal hospital with content and images that explain physical examination, radiology, surgery, and drug therapy for diseases and conditions common to dogs and cats. The poison information covers the dangers of lawn and garden products (fertilizers, pesticides), flea control products, household products (medications, mothballs, batteries, cigarettes, etc.), and more.

Toxic Products—Dog Guide
<http://animal.discovery.com/guides/dogs/afety/toxicproducts.html>

Animal Planet's dog guide lists household products that are toxic to dogs including cleaning fluids, solvents, over-the-counter medications (pain relievers, laxatives, diet pills), shampoos, deodorants, chocolate, matches, furniture polish, shoe polish, and similar products.

SERVICE ANIMALS

The term service animal refers to animals that are trained to assist persons with various hearing, sight, mobility, or balance impairments. In some cases, a service animal is used to provide seizure alerts. Although the term usually refers to trained dogs, other animals such as monkeys, pigs, and cats have been trained to assist the disabled. See also Therapy Pets featured later in this chapter for animals trained to provide solace, companionship, or entertainment to hospitalized, convalescing, or elderly persons.

☑ Helping Hands—Monkey Helpers for the Disabled
<http://www.helpinghandsmonkeys.org/>

Monkey Helpers for the Disabled, Inc. is a nonprofit organization headquartered in Boston, Massachusetts that established Monkey College to train Capuchin monkeys to work with disabled persons, especially those who have been paralyzed by spinal cord injuries. Monkeys have been placed with persons who have become impaired for a variety of reasons, focusing on "people who have become quadriplegic (paralyzed from the neck down) as a result of an accident, injury, or disease." Monkeys are trained to assist with various activities of daily living such as "turning on a light, picking up a dropped or out-of-reach object, or turning the pages of a book."

Service Animal Information
<http://www.usdoj.gov/crt/ada/animal.htm>

The U.S. Department of Justice (DOJ) uses the Americans with Disabilities Act (ADA) definition of service animal and answers commonly asked questions about allowing persons with service animals in areas (such as businesses) normally restricted to just humans.

Should They Stay or Should They Go?
<http://www.vetcentric.com/magazine/magazineArticle.cfm?ARTICLEID=1792>

VetCentric is a site that focuses on the health and behavior of dogs, cats, horses, birds, fish, ferrets, rabbits, "pocket pets," reptiles, farm animals, and wildlife. The article discusses the pros and cons of taking a pet animal along when traveling.

Tips for People with Service Animals or Pets
<http://www.prepare.org/disabilities/animaltips.htm>

The American Red Cross's prepare.org site offers sound and practical advice for persons with service or pet animals who need to evacuate an area. Tips include preplanning, equipping an emergency kit (bedding, seven days' worth of food, water/food bowl, plastic bags, paper towels, a favorite toy, antibacterial ointment, extra harness), and more.

Travel with a Service Animal
<http://www.gimponthego.com/tips2.htm>

The Gimp on the Go site includes useful tips for handicapped persons traveling with service animals. Information includes air travel and hotel accommodations.

SPAYING AND NEUTERING

According to SPAY/USA <http://www.spayusa.org/>, a single, unspayed female cat, her partner, and her offspring can yield more than 11 million kittens over nine years. Pet sterilization also helps with various behavioral problems such as spraying, marking, and aggressive behavior. Spaying refers to the sterilization of female animals; neutering is the term used for the sterilization of male animals.

Myths and Facts About Spaying and Neutering
<http://www.hsus.org/pets/pet_care/myths_and_facts_about_spaying_and_neutering.html>

The Humane Society of the United States (HSUS) focuses on the humane and compassionate treatment of pets, wildlife, farm animals, laboratory animals, and marine mammals. The information about spaying and neutering covers common myths and facts about the subject such as weight gain, the need for female animals to have at least one litter, the expense of the procedure, and other misconceptions.

Responsible Pet Care
<http://www.homeagainid.com/health/rpc/spayneuter.cfm>

HomeAgainID.com is a company that specializes in microchipping of pets. The site focuses on three aspects of responsible pet ownership: spaying/neutering, vaccinations, and microchipping. The information about spaying/neutering includes overall health, contentedness, costs of the procedure versus costs of a litter, and problems of pet overpopulation.

SPAY/USA
<http://www.spayusa.org/>

SPAY/USA is a program of the North Shore Animal League in Port Washington, New York. The organization's site features frequently asked questions, dictionary of terms, benefits of spaying/neutering cats or dogs, statistics related to pet population and euthanized pets, sterilization methods, legislation, and efforts to spay/neuter feral cats.

Will Neutering Change Your Pet's Personality?
<http://www.vetcentric.com/magazine/magazineArticle.cfm?ARTICLEID=1372>

VetCentric is a site that focuses on the health and behavior of dogs, cats, horses, birds, fish, ferrets, rabbits, "pocket pets," reptiles, farm animals, and wildlife. This article focuses on the common misconception that spaying/ neutering will change a pet's personality.

SURGERY, SURGICAL PROCEDURES, AND HOSPITALIZATION

Behavior Problems Following Hospitalization
<http://www.bobmckee.com/Client%20Info/Behaviour/Hospitalization.html>

Veterinarian Bob McKee explains behavior problems that may result after a hospitalized pet returns home including possible depression, aggressive behavior with other pets in the household, and inappropriate elimination.

Post-Op Care for Pets
<http://www.vetcentric.com/magazine/magazineArticle.cfm?ARTICLEID=1889>

VetCentric is a site that focuses on the health and behavior of dogs, cats, horses, birds, fish, ferrets, rabbits, "pocket pets," reptiles, farm animals, and wildlife. The information about postoperative care covers the importance of discharge instructions, pain relief, and good nutrition, and being vigilant about wound care, constipation, and scratching.

Stages of Anesthesia in Animals
<http://www.peteducation.com/article.cfm?cls=0&cat=1478&articleid=979>

Veterinarians Foster and Smith's information about how pets handle anesthesia covers the use of sedatives prior to induction of anesthesia, keeping the pet unconscious and pain-free during the procedure, and postoperative care.

Veterinary Hospital Locator
<http://www.healthypet.com/hospital_search.aspx>

The Healthypet.com site features a database of animal hospitals that is searchable by state/province, ZIP code, or name, within one to 100 miles.

The hospitals that are included in the database belong to the American Animal Hospital Association (AAHA).

Veterinary Surgery
<http://www.peteducation.com/category_summary.cfm?cls=0&cat=1477>

Veterinarians Foster and Smith's information about animal surgery procedures covers diagnostic biopsies, anesthesia planning, therapeutic procedures, and more.

THERAPY PETS

Animal-assisted therapy was developed more than forty years ago in response to the realization that companion animals bring great comfort to the ill and infirmed.[2] Therapy animals serve a different function than service animals but still undergo rigorous training and testing for temperament and obedience. See also the section on Service Animals.

Caring Canines
<http://www.flagstaffmedicalcenter.com/pp_fmc/dept_services/caring_canines_fmc.htm>

Flagstaff Medical Center in Flagstaff, Arizona developed Caring Canines, an animal-assisted therapy program for inpatients. This site explains the steps that pet owners need to take to get their dogs certified as therapy pets with the Delta Society's Pet Partners program.

Delta Society
<http://www.deltasociety.org/>

Delta Society is dedicated to training animals to provide animal-assisted therapy in a variety of settings. Site features include articles about the health benefits of companion animals, information about Pet Partners workshops and evaluative testing for potential therapy pets, resources for pet loss and bereavement, and information about service animals. Some site content is restricted to Delta Society members (donors who contribute $50 or more per year), and Pet Partners evaluators and instructors.

Love on a Leash
<http://www.loveonaleash.org/>

Liz Palika, a dog obedience instructor and author, founded "Love on a Leash," a nonprofit organization headquartered in Oceanside, California. The organization's mission is to "provide training, evaluation, and certification procedures for therapy pets and their owners." Therapy pets include dogs, cats, and other animals.

Pet Therapy
<http://www.sniksnak.com/therapy.html>

Pawprints and Purrs, Inc. is a Louisiana-based nonprofit organization that concerns itself with the welfare and well-being of cats and dogs. The information about pet therapy includes examples of pets healing their owners, and helping other people with various conditions and diseases.

Service and Therapy Dogs
<http://www.cofc.edu/~huntc/service.html>

The College of Charleston's Caroline Hunt is an English professor and dog trainer. Her site explains the difference between a service animal and a therapy pet, includes her online lecture about therapy dogs, recommends links for persons interested in service dogs or therapy dogs, and more.

Therapy Dogs—Getting Started
<http://www.amchessieclub.org/TherapyDog.html>

The American Chesapeake Club site explains the desired traits for therapy dogs, based on a modification of the American Kennel Club's Canine Good Citizenship test, and provides answers to frequently asked questions (age of dog, expectations of handlers, registering with therapy dog organizations, etc.).

Therapy Dogs International (TDI)
<http://www.tdi-dog.org/>

TDI, headquartered in Flanders, New Jersey, has registered more than 14,000 dogs to visit residents and patients in nursing homes, hospitals, and other institutions. The TDI site explains the details of the test used to evaluate potential therapy dogs. One of TDI's newest services is its "Children Reading to Dogs" program.

Visiting Pets and Animal-Assisted Therapy
<http://www.dog-play.com/therapy.html>

DogPlay was developed by Diane Blackman to sell playful dog-related products and promote dog play. The information about therapy pets gives an excellent overview of animal-assisted therapy, groups that provide training and evaluation of therapy dogs, the traits of a dog temperamentally suited to therapy work, organizations that test and register therapy animals, and more. The site also sells bibs, T-shirts, jersies, sweatshirts, bumper stickers, etc. of interest to persons involved in rescue, therapy, agility, herding, etc.

TRAVELING WITH PETS AND PETS TRAVELING ALONE

This section includes links for persons who wish to travel with their pets by air, bus, train, or car, or ship their pets to different locations. Depending on the destination, season of the year, and other factors such as pet breed or size, airlines may impose restrictions and certain countries have restrictive quarantines.

Air Animal
<http://www.airanimal.com/>

Air Animal is a pet moving company based in Tampa, Florida that was established in 1977 by a veterinarian and his wife. Their company's site emphasizes the importance of rabies vaccination history documentation, preplanning for international relocations, EuroChip for pet identification in European Union countries, pet travel health certificates, and more.

☑ **Air Travel with Dogs**
<http://www.akc.org/pdfs/canine_legislation/airline_chart_0406.pdf>

The American Kennel Club (AKC) developed this handy chart to explain specific airline policies, fees, and restrictions related to shipping dogs or traveling with dogs.

Canine Travel Tips
<http://www.akc.org/public_education/travel.cfm>

The American Kennel Club (AKC) offers practical advice for persons planning to travel by car, plane, bus, boat, or train with dogs or ship dogs domestically or internationally.

Caring for Pets When You Travel
<http://www.hsus.org/pets/pet_care/caring_for_pets_when_you_travel/>

The Humane Society of the United States (HSUS) focuses on the humane and compassionate treatment of pets, wildlife, farm animals, laboratory animals, and marine mammals. The site also includes information about pretrip preparation, pet-friendly accommodations, travel supplies, and traveling by air, car, ship, or train.

☑ DogFriendly.com
<http://www.dogfriendly.com/>

This site features city guides and travel guides for dog/pet owners wishing to travel throughout the United States and Canada. The site also links to off-leash dog parks, recreational vehicle (RV) parks and campgrounds, beaches, and other dog-friendly attractions (Figure 3.3).

Dogpark.com
<http://www.dogpark.com/>

Dogpark.com is a San Rafael, California-based site that focuses on off-leash fun for dogs in California and parts of Canada. Site information includes some dog-friendly accommodations, car or air travel, dog treat recipes, grooming, Digital Dog Park discussion board, and a twenty-two-page list of dog parks from Arizona to Washington State.

FlyPets.com
<http://www.flypets.com/>

Pet Air is a company that specializes in animal transportation by air. Its FlyPets site includes price estimates, specific airports, online reservations, supplies, and special services (door-to-door, red-eye flights, etc.).

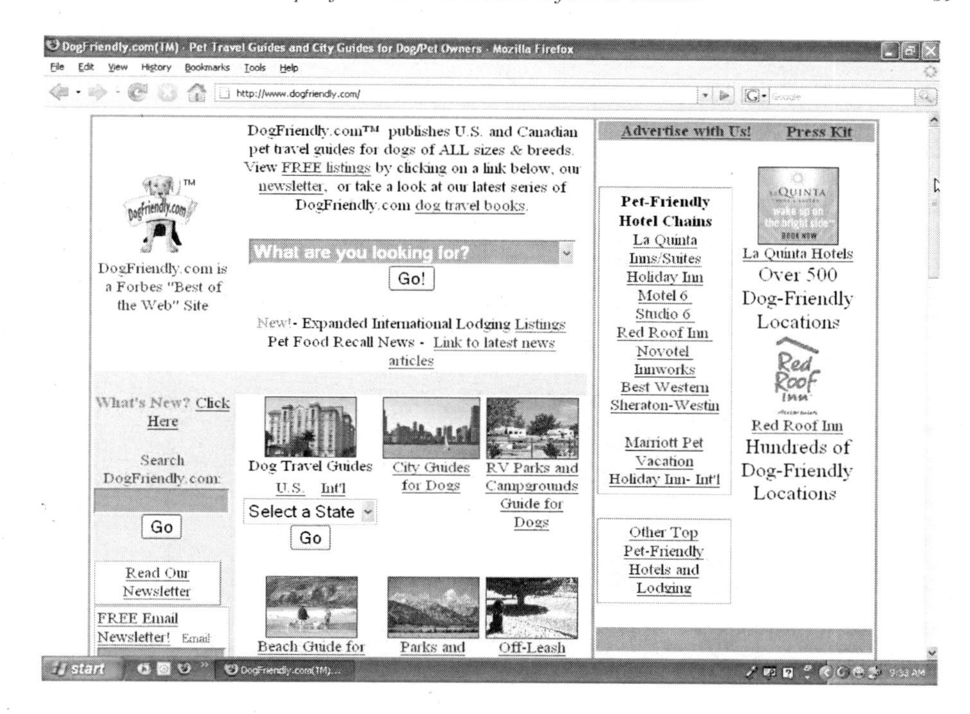

FIGURE 3.3. DogFriendly.com
<http://www.dogfriendly.com/>
Reprinted with permission.

☑ Happy Tails Travel <http://www.happytailstravel.com/>

Happy Tails Travel is a Tucson, Arizona-based company that focuses on pet travel and shipping services. Site features include information about pet shipping as a tax-deductible expense, pet sitting, local pet transportation in southern Arizona, pet relocation for military personnel, pet relocation planning and cost estimates, and more (Figure 3.4).

☑ Importation of Pets <http://www.cdc.gov/ncidod/dq/animal.htm>

The U.S. Centers for Disease Control and Prevention (CDC) is concerned with the monitoring and prevention of disease. Site information about im-

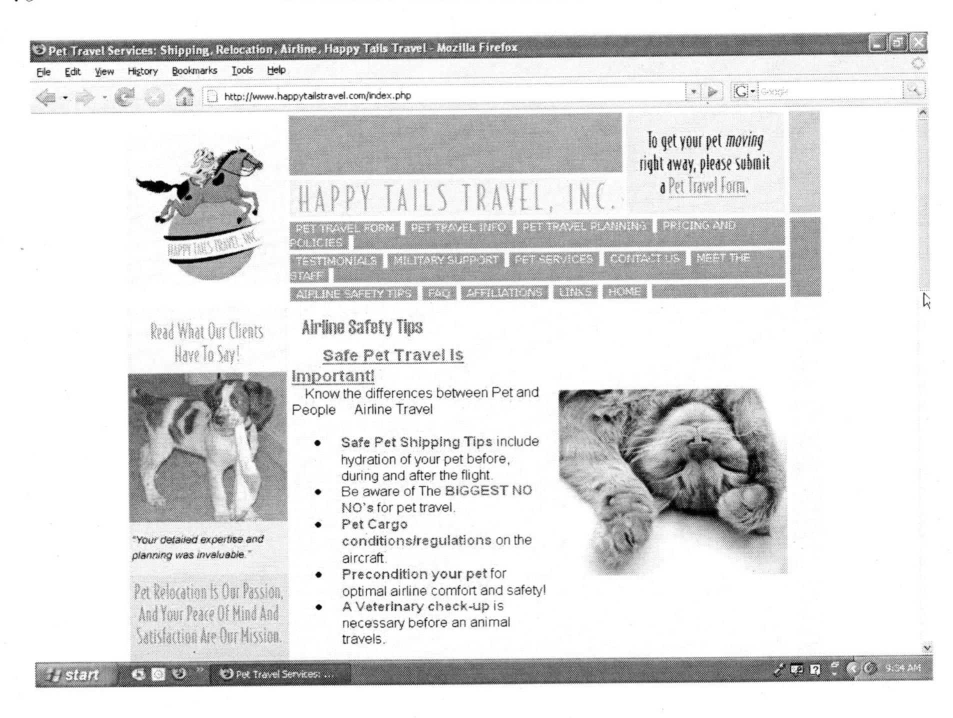

FIGURE 3.4. Happy Tails Travel <http://www.happytailstravel.com/> Reprinted with permission.

porting pets into the United States and its territories includes embargo news, animal import regulations, restricted animals, and more (Figure 3.5).

Pet Emergencies <http://www.bringyourpet.com/emergencies.htm>

BringYourPet.com is a site that focuses on pet-friendly lodging available in the United States, Canada, and other parts of the world. The information about preparing for pet emergencies while away from home features a list of first aid supplies, emergency contact numbers, and deciding whether to consult a veterinarian.

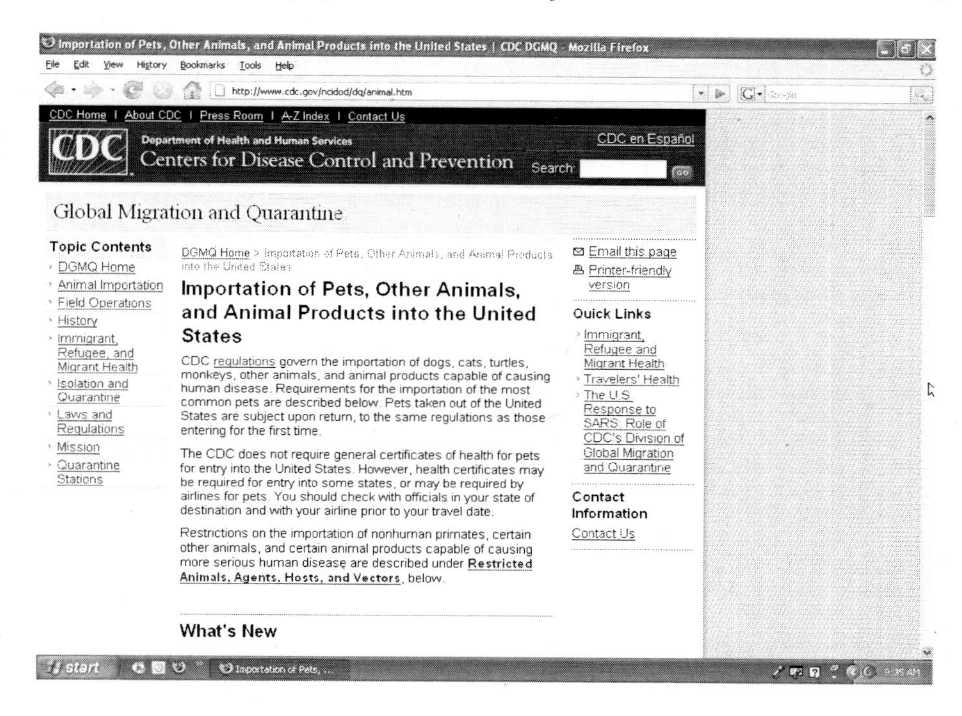

FIGURE 3.5. Importation of Pets
<http://www.cdc.gov/ncidod/dq/animal.htm>

Pet Taxi
<http://www.epettaxi.com/>

Pet Taxi is a New York-based company that offers national and international pet shipping services including quarantine, vaccinations, and other health requirements.

☑ Pets on the Go!
<http://www.petsonthego.com/>

Pets on the Go! is a site that focuses on products and services for people who like to travel with their pets. Site features include information about pet-friendly accommodations (motels, hotels, resorts, and rental properties), pet-friendly transportation (car rentals, air travel, and cruise companies), pet movers, pet sitters, and more.

☑ Petswelcome
<http://www.petswelcome.com/>

This well-designed site includes pet-friendly lodging listings in the United States and Canada; travel tips related to traveling by air; relocating with pets; pet regulations by foreign country; disaster planning and evacuation hints; and discussion forums.

Shipping Live Animals
<http://www.hawaiianair.com/cargo/pets/>

Hawaiian Airlines' information about shipping live animals (in the cabin, as checked baggage, or as cargo) includes useful tips applicable to other situations including quarantine, temperature restrictions, approved kennel dimensions, and restricted breeds.

Takeyourpet.com
<http://www.takeyourpet.com/>

Although this site has a $1.95 membership fee for accessing its pet-friendly lodging listings and receiving deep discounts, much of the remaining information is available for free, including its useful tips about pretravel checkups, traveling by car or plane, lodging etiquette, travel supplies, and more. The site's free e-mail newsletter features travel news and tips.

TravelDog.com
<http://www.traveldog.com/>

TravelDog.com offers helpful information about dog-friendly accommodations, destinations, and pet etiquette. Site membership is $9.95 per year, which offers discounts on lodging, products, and services. Notable features include the members-only discussion board and dating service for single pet lovers.

Traveling with or Boarding Your Dog
<http://www.petplace.com/browse.aspx?a=1&c=577>

Veterinarian Jon Rappaport's information about traveling with or boarding dogs concentrates on airline rules, advantages and disadvantages of using sedatives, boarding the dog at a kennel facility, hiring a pet sitter, motion sickness while traveling, etiquette when staying in a hotel with a dog, and much more.

Traveling with Pets
<http://www.petstyle.com/lifestyle_article.aspx?section=travel>

PetStyle's site information about pet travel covers pet-friendly airlines, pet-friendly hotels, crates, kennels, shipping pets, moving with a pet, pet sitters, and more.

Traveling with Pets
<http://seniors-site.com/travel/pets.html>

Seniors-Site.com, a site that offers useful information for adults who are fifty years of age and older, includes succinct checklists for traveling with pets by plane or car.

VACCINATIONS AND IMMUNIZATIONS

Vaccinations and immunizations protect pet animals and humans alike, but there are some controversies about their frequency, effectiveness, and adverse effects, sometimes called vaccinosis.

Canine Vaccinations Full Report
<http://www.aahanet.org/graphics/pdf/Canine%20Vaccine%20FULL%20REPORT.pdf>

The American Animal Hospital Association (AAHA) developed this twenty-eight-page report related to routine vaccinations for dogs. Veterinarians are not required to follow these 2003 guidelines. Highlights include age, dosage, and booster frequency for canine distemper virus, canine parvovirus, canine adenovirus-1, canine adenovirus-2, rabies virus, distemper-measles virus, parainfluenza virus, *Leptospira interrogans* (combined with serovars canicola and icterohaemorrhagiae), *Bordetella bronchiseptica, Borrelia burgdorferi,* and *Giardia lamblia* vaccines. The developers of these guidelines also explain the duration of immunity for various canine vaccine antigens. AAHA developed a similar document related to feline vaccine controversies but access is restricted to AAHA members.

Cat Diseases and Vaccines <http://www.shelterpetsink.com/cat_diseases.htm>

Shelter Pets Ink is a site that focuses on "changing animals' lives through writing." The information about the link between routine vaccinations and cat diseases was written by Kari Winters, a psychiatric nurse who is dedicated to animal rescue.

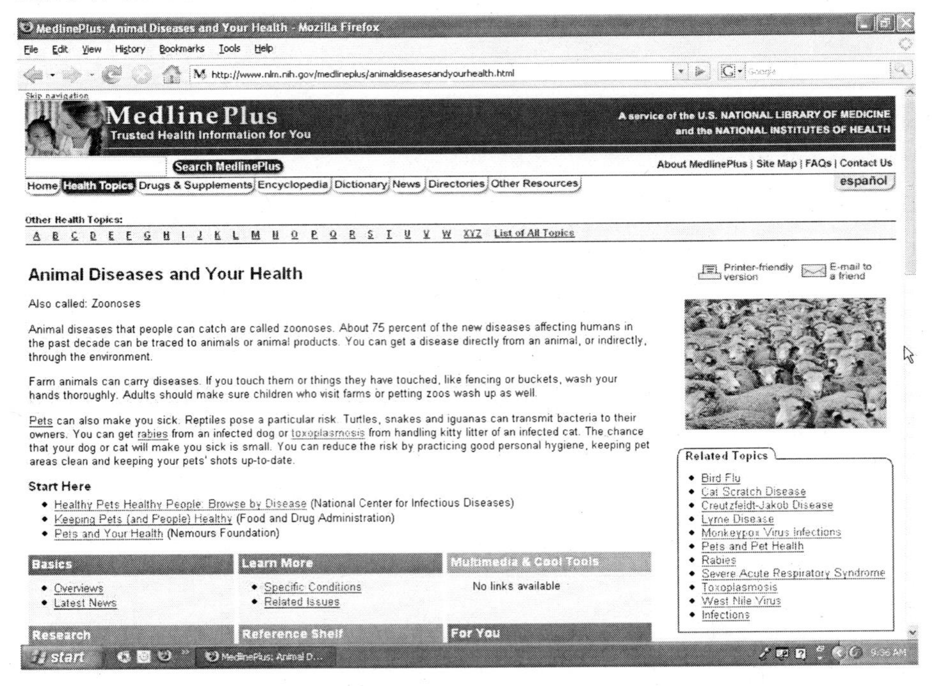

FIGURE 3.6. Animal Diseases and Your Health <http://www.nlm.nih.gov/medlineplus/animaldiseasesandyourhealth.html>

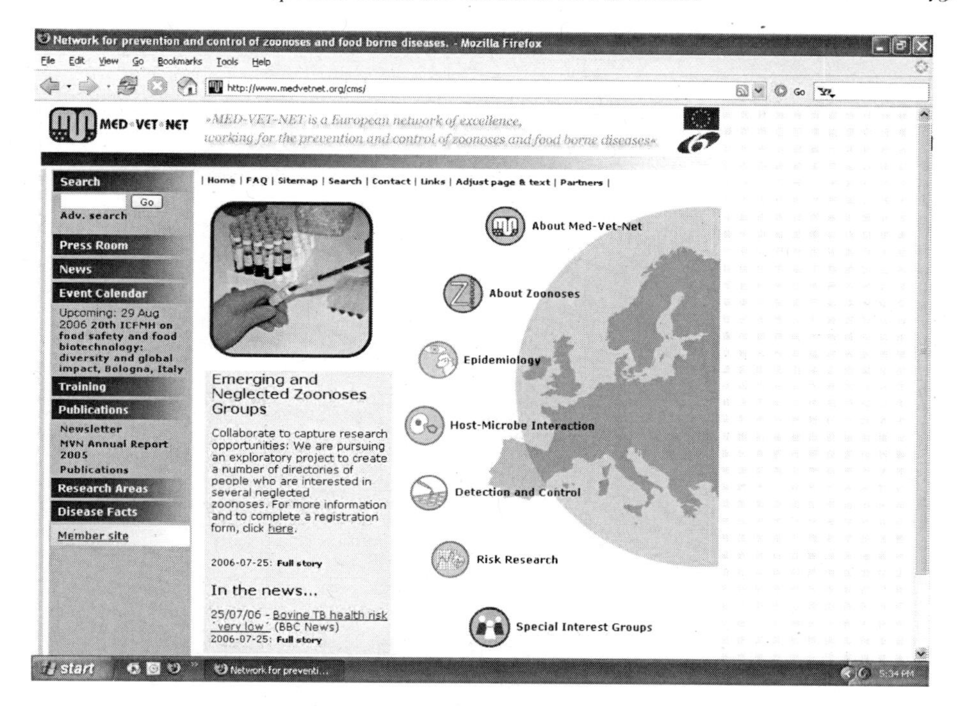

FIGURE 3.7. MED-VET-NET
<http://www.medvetnet.org/>
Reprinted with permission.

☑ Feline Vaccination Guidelines <http://www.aafponline.org/resources/guidelines/2006_Vaccination_Guidelines_JAVMA_%20PDF_Plus.pdf>

The American Association of Feline Practitioners (AAFP) and Academy of Feline Medicine Advisory Panel on Feline Vaccines developed this twenty-nine-page report to advise veterinarians on the routine administration of feline panleukopenia, feline viral rhinotracheitis, feline calcivirus infections, rabies virus, feline leukemia virus infection, chlamydiosis, feline infectious peritonitis, dermatophytosis, *Bordetella bronchiseptica* infection, and Giardiasis vaccines to kittens and cats. The document also discusses liability issues, vaccine labels, reporting of adverse effects, and more.

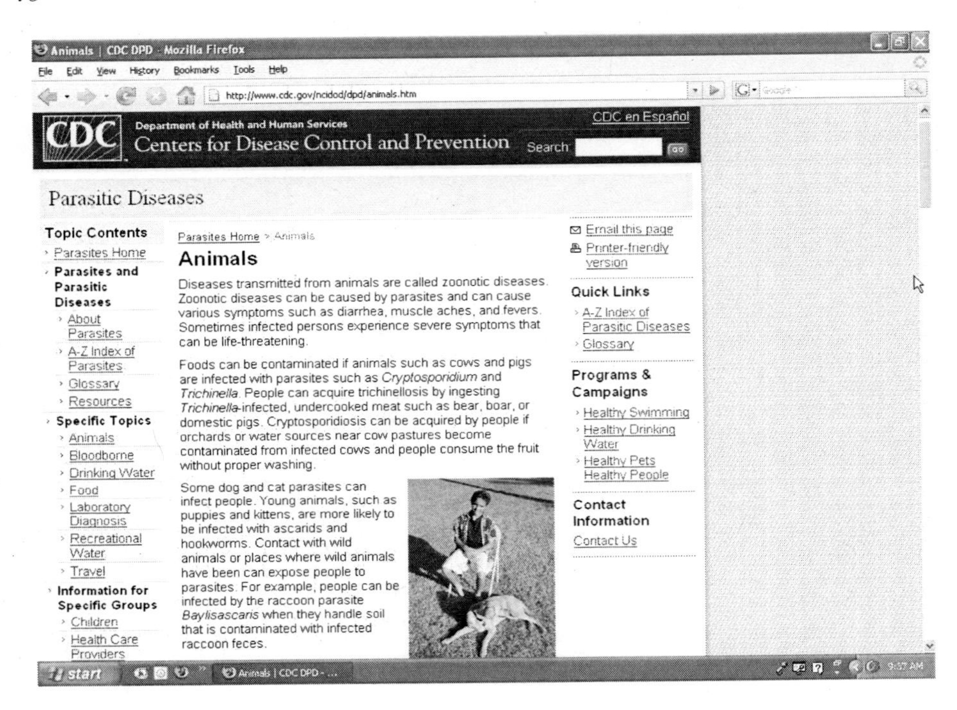

FIGURE 3.8. Parasitic Diseases—Animals <http://www.cdc.gov/ncidod/dpd/animals.htm> Reprinted with permission.

Preventative Care for Pets <http://www.vet4petz.com/index_prevention.htm>

Vet4Petz is a site that offers useful pet health information for site members and nonmembers alike. Site features include signs of a medical emergency, feature articles, health tips and tidbits, preventive screenings and vaccinations, alternative medical treatments, pet pharmacy, and pet products. Site members pay $19.95 per year, which entitles them to "to our online database of Pet Health Questions and Answers from a Licensed Veterinarian." The information about pet diagnostic tests and vaccinations covers annual physical examinations, feline leukemia and feline AIDS testing, blood tests for older pets, heartworm tests, stool samples, teeth cleaning, nail trimming, and various vaccines (rabies, Lyme disease, distemper, parvovirus, feline infectious peritonitis, etc.).

Understanding Vaccines
<http://www.cahi-icsa.ca/comm-factsheets-vaccines.php>

Canadian Animal Health Institute (Institut Canadien de la Sante Animale) provides information about types of biological substances that impart disease immunity, precautions to take to reduce side effects, and how vaccines work.

Vaccinating Your Dog
<http://www.healthypet.com/library_view.aspx?ID=196&sid=1>

HealthyPet is a site developed and maintained by the American Animal Hospital Association (AAHA). The information about vaccinations for dogs mentions their importance for preventing serious illnesses, possible side effects, core vaccines (distemper, adenovirus, parvovirus, and rabies) and noncore vaccines that are recommended for exposed pets (kennel cough, Lyme disease, leptospirosis), and frequency of vaccine boosters.

ZOONOTIC DISEASES

Zoonotic diseases are transmitted from animals to humans and can pose serious health risks to anyone who comes into contact with an infected animal, or its bodily fluids. The most common health risks for pet owners include toxoplasmosis, rabies, cat scratch disease, ringworm, and scabies.

Animal Diseases and Your Health
<http://www.nlm.nih.gov/medlineplus/animaldiseasesandyourhealth.html>

MedlinePlus is a consumer health resource developed by the U.S. National Library of Medicine that provides extensive information about more than 650 diseases and conditions. The information includes overviews of animal diseases that can harm humans, and specific diseases transmitted by birds (salmonellosis, psittacosis, cryptococcosis, etc.), cats (cat scratch diseases, toxoplasmosis, campylobacteriosis, etc.), dogs (brucellosis, tapeworm, giardiasis, etc.), fishes and amphibians (salmonellosis, and risks to immunocompromised persons), pocket pets (lymphocytic choriomeningitis, salmonellosis, etc.), reptiles (salmonellosis), and more (Figure 3.6).

Bartonella and Cat Scratch Disease
<http://www.marvistavet.com/html/bartonella.html>

The Mar Vista Animal Medical Center in Los Angeles, California offers a wide range of pet health information on their site. Caused by the *Bartonella henselae* organism, cat scratch disease is spread by fleas and transmitted from an infested cat to another cat or a person.

Healthy Pets Healthy People
<http://www.cdc.gov/healthypets/browse_by_animal.htm>

The U.S. Centers for Disease Control and Prevention (CDC) is concerned with the monitoring and prevention of disease. The information about diseases transmissible from animals to humans is searchable by type of animal (birds, cats, dogs, farm animals, fish, horses, reptiles, wildlife, and pocket pets).

☑ MED-VET-NET
<http://www.medvetnet.org/>

MED-VET-NET describes itself as "a European network of excellence, working for the prevention and control of zoonoses and food-borne diseases." The site explains three types of diseases that are transmissible from animal to human: bacterial, viral, and parasitic; the frequency and control of these diseases; and risk research (Figure 3.7).

☑ Parasitic Diseases—Animals
<http://www.cdc.gov/ncidod/dpd/animals.htm>

The U.S. Centers for Disease Control and Prevention (CDC) is concerned with the monitoring and prevention of disease. Information about diseases transmitted from animals to humans includes avoiding contact with wild animals, safeguards for HIV-positive individuals, toxoplasmosis, and more (Figure 3.8).

Vector Borne and Zoonotic Disease Section
<http://www.azdhs.gov/phs/oids/vector>

The Vector Borne and Zoonotic Disease Section of the Arizona Department of Health Services "is responsible for monitoring, controlling, and preventing diseases transmitted from animals or arthropod vectors to humans," by monitoring anthrax, viral meningitis, brucellosis, Colorado tick fever, Creutzfeldt-Jakob disease, cysticercosis, dengue fever, hantavirus, leptospirosis, lymphocytic choriomeningitis, malaria, plague, psittacosis, Q-fever, rabies, borreliosis, Rocky Mountain spotted fever, tularemia, West Nile virus, yellow fever, and other diseases. Site features include a newsletter, how rabies is monitored in the United States, statistical data, and more.

Zoonoses Guidelines
<http://www.aafponline.org/resources/guidelines/ZooFinal2003.pdf>

American Association of Feline Practitioners (AAFP) is a professional organization of veterinarians who practice feline medicine and surgery. AAFP's thirty-two-page guidelines related to disease transmitted by animals covers intestinal organisms (worms, protozoa, bacteria, etc.), bites and scratches, respiratory exposure, skin exposure, cats as "sentinels of human exposure" for bioterrorist attacks, and much more.

Zoonoses—Introduction
<http://www.merckvetmanual.com/mvm/index.jsp?cfile=htm/bc/220100.htm&word=zoonoses>

Drug manufacturer Merck & Co., Inc.'s veterinary manual includes more than 12,000 topics including an introduction to diseases that can be transmitted from animals to humans (zoonotic diseases), and the susceptibility of immunocompromised people to mycobacterial infections, toxoplasmosis, cryptosporidial enteritis, Salmonella infections, and others.

☑ Zoonoses—World Health Organization
<http://www.who.int/zoonoses/>
<http://www.who.int/topics/zoonoses/en>

The World Health Organization (WHO) is the health agency of the United Nations that concerns itself with the "physical, mental, and social well-being" of the world's populations. The information about diseases that can be spread from animals to humans includes outbreak alerts, specific diseases (anthrax, animal influenza, bovine spongiform encephalopathy, brucellosis, animal diseases spread by foods and food products, hemorrhagic fevers, echinococcosis, leishmaniasis, leptospirosis, prion diseases, rabies, cysticercosis, trematodosis, trypanosomiasis, and tularemia), and documents related to many of these disease threats.

Zoonoses Manual
<http://www.lapublichealth.org/vet/guides/vetzooman.htm>

The Los Angeles (California) Department of Public Health organized this listing of reportable diseases including ones that are spread by animals, reverse zoonotic diseases—meaning diseases that can be spread from humans to animals (mumps, hepatitis, diphtheria, furuncolosis, scarlet fever, giardiasis, and tuberculosis), and more.

NOTES

1. *Pet Insurance in North America: The Market and Trends in the U.S. and Canada.* New York: MarketResearch.com, Inc. 2005.
2. Hooker, Shirley D., Freeman, Linda Holbrook, and Stewart, Pamela. (2002). "Pet Therapy Research: A Historical Review." *Holistic Nursing Practice,* 17(1): 17-23.

Chapter 4

For Cats Only

Dogs come when they're called; cats take a message and get back to you.

Mary Bly

A cat's rage is beautiful, burning with pure cat flame, all its hair standing up and crackling blue sparks, eyes blazing and sputtering.

William S. Burroughs

Cats are smarter than dogs. You can't get eight cats to pull a sled through snow.

Jeff Valdez

GENERAL CAT HEALTH SITES

☑ Cat Health Center
<http://animal.discovery.com/guides/healthcenter/cats/cats.html>

Animal Planet features a cat guide that is searchable by part of the body or body function, choosing a cat, cat care, cat behavior, training, safety (plants to avoid, first aid, multiple cats in the same household), and more.

Cat Health Information by Condition or Disease
<http://www.sniksnak.com/cathealth/>

Pawprints and Purrs, Inc. is a Louisiana-based nonprofit organization that concerns itself with the welfare and well-being of cats and dogs. The section devoted exclusively to cats includes an alphabetic listing of

Internet Guide to Pet Health

doi:10.1300/5675_04

diseases and conditions from aggression to worms. A typical entry includes an understandable disease explanation, recommendations, and tips.

CatHobbyist
<http://forums.cathobbyist.com/forum.php?catid=21>

CatHobbyist is an online community of cat lovers. Site features include threaded discussions about cat health, photo gallery, and external links on a variety of subjects such as rescue/adoption, societies, and pet stores. Although the site is cluttered with links to various commercial sites, cat lovers will appreciate the friendly advice dispensed here as well as the location-specific resources. Similar sites exist for owners of dogs (DogHobbyist .com), reptiles (Kingsnake.com), insects (InsectHobbyist.com), birds (Bird Hobbyist.com), fish (AquariumHobbyist.com), and all pets (PetHobbyist .com).

TheCatSite.com
<http://www.thecatsite.com/>

Founded by cat behaviorist and writer Anne Moss, TheCatSite.com focuses on cat behavior (cat fights, urine spraying, meowing), cat care (hairballs, household chemicals, holiday dangers), cat forums, cat health (ear mites, diabetes, allergies, etc.), cat shop (popular cat toys, furniture, flea medication, etc.), cat breeds, and more.

☑ Feline Advisory Bureau (FAB)
<http://www.fabcats.org/>

FAB is a U.K.-based charitable organization that is "dedicated to promoting the health and welfare of cats through improved feline knowledge." Site features include information for cat owners (such as cat case histories, tributes to departed pets, health concerns from anemia to hyperthyroidism to cat diseases transmissible to humans, cat behavior topics from aggression to catnip to wool eating, and more), information for breeders, information about becoming a veterinary nurse, and more.

☑ Indoor Cat Initiative
<http://www.vet.ohio-state.edu/indoorcat>

The College of Veterinary Medicine at Ohio State University and the Kenneth A. Scott Charitable Trust helped fund this rich resource for owners of indoor cats. Site information includes creating an ideal home environment, unique features that affect a cat's wake-sleep cycle, supplies and toys, life stressors for cats (such as visiting the veterinarian, introduction of a new pet, household construction, traveling, hospitalization, etc.), problem solving (spraying and marking, new pets, separation anxiety, cat conflicts, physical activity), and much more.

Messybeast.com Cat Resource Archive
<http://messybeast.com/catarchive.htm>

This site describes itself as a "not-for-profit archive of information on cat care, welfare, behavior, breeding, rescue, and feline general interest." Site features include annotated links on a variety of subjects including neutering, euthanasia, aging, predatory activity, and much more.

Purina ONE—Cat Conditions A to Z
<http://www.purinaone.com/catcare_cond_atoz.asp>

Purina ONE is a brand of pet food. The information about cat care includes a library of articles about adverse effects of vaccinations, weight problems, kitten growth, and development, signs of a healthy cat, wellness quizzes, breed selector, and more. Site visitors can register for an e-mail newsletter that focuses on product news and advice for cat owners.

What About Cats?
<http://www.whataboutcats.com/diseases.html>

This site is a delightful collection of text and images related to first aid; household dangers; hazardous indoor, outdoor, and seasonal plants; cat animations developed by Lucy Rand and other artists; list of pet grief hotlines; information about various infectious diseases; and more.

SPECIFIC DISEASES AND CONDITIONS IN CATS

Cats can suffer from a variety of diseases and conditions, some of which are chronic and life threatening. Some breeds are predisposed to certain maladies and conditions, and other conditions are related to age.

BEHAVIORAL PROBLEMS

Some of the leading causes for animal abandonment are behavioral problems such as excessive vocalization, inappropriate elimination, and/or aggression toward humans. Exercise, training, environmental improvements, and play can be used to modify unwanted behaviors in cats.

☑ The Behavior Clinic
<http://www.tufts.edu/vet/behavior/feline.shtml>

The Behavior Clinic at Tufts University's Cummings School of Veterinary Medicine focuses on the understanding and treatment of various behavioral problems in cats including aggression, anxiety, urine spraying, wool sucking, fighting among cats, and more.

Cat Behavior
<http://www.ársscats.com/Cat_Behavior.htm>

PussCats.com is a site that focuses on subjects of interest to cat lovers and cat owners. The information about cat behavior ranges from body language (head, ears, whiskers, and tail), meowing, spraying and marking, aggression, and training.

Cat Channel
<http://www.catchannel.com/>

Cat Channel is a site that provides useful cat health content from magazines published by BowTie. Inc., (*CAT FANCY, Cats USA, Cats Magazine,* and *Kittens USA*). Site highlights include breed profiles, topics related to behavior (aggression, inappropriate elimination, indoor cats, play activities, etc.), health (dental care, senior cats, grooming, diet and nutrition, travel, etc.), activities, health problems (allergies, digestive problems, feline diabetes, parasites, respiratory diseases, skin, and coat), quizzes, contests, cat discussion boards, and more.

Cats—Behavior
<http://www.petplace.com/cat-behavior.aspx>

Veterinarian Jon Rappaport's site includes information about cat behavior covers medical causes of aggression, cognitive dysfunction in aging cats, hypervocalization, urine marking, hoarding behavior, aggression toward other cats in the household, scratching, and much more.

Cats International
<http://www.catsinternational.org/>

Cats International is a nonprofit organization based in Cedarburg, Wisconsin that is "dedicated to helping people better understand their feline companions." Site features include articles (house soiling, aggression, scratching and clawing, natural cat behavior, benefits of spaying/neutering, etc.) relating to cat behavior.

***The Daily Cat*—Behavior Basics**
<http://www.thedailycat.com/Behavior/index.html/>

The Daily Cat is an online magazine devoted to understanding cats and their behavior. The section related to behavior problems covers inappropriate urination and defecation, scratching, biting, cognitive disorders, and more.

Feline Behavior Problems
<http://www.newmanveterinary.com/feline.html>

Newman Veterinary Medical Services in Washington State hosts a site that features useful pet health content developed by veterinarian Newman. Behavior problems include predatory aggression, fear, inappropriate elimination, furniture scratching, and more.

☑ Feline Behavior Problems (AAFP)
<http://www.aafponline.org/resources/guidelines/Feline_Behavior_Guidelines.pdf>

American Association of Feline Practitioners (AAFP) is a professional organization of veterinarians who practice feline medicine and surgery. This forty-four-page document describes and explains normal behavior in cats, predatory behavior, aggression among cats, normal elimination, how cats learn, harmful stress that can result in behavior problems, treatments (drug therapy, behavior modification), and much more. The document also includes useful handouts for veterinary customers on a variety of topics including enriching the quality of a cat's life, reducing elimination problems by taking scrupulous care of the litterbox, adopting a new cat, and more.

Feline House Soiling
<http://www.marvistavet.com/html/feline_house_soiling.html>

The Mar Vista Animal Medical Center in Los Angeles, California offers a wide range of pet health information on their site. Site content related to why cats urinate or defecate outside the litterbox covers possible medical causes (urinary tract disease), urine marking, and drug treatments.

Preventing Behavior Problems in Your Cat
<http://www.bestfriends.org/theanimals/pdfs/cats/preventingbehaviorproblems.pdf>

While working as a resident at the University of California at Davis School of Veterinary Medicine, veterinarian Sheila Segurson developed this two-page document that explains soiling, aggression, and scratching, and some remedies (scooping the litterbox daily, trimming nails, providing surfaces, training with clickers) that may work.

Separation Anxiety in Cats
<http://www.peteducation.com/article.cfm?cls=1&cat=1310&articleid=2372>

Veterinarians Foster and Smith's site features this article by veterinarian Katharine Hillestad about how separation anxiety in cats can be remedied through behavior modification and antianxiety drugs (Buspar, Prozac, or Clomicalm).

BLADDER, KIDNEYS, AND URINARY TRACT

Bladder and Kidney Stones in Dogs and Cats
<http://www.2ndchance.info/calculi.htm>

Veterinarian Ron Hines of All Creatures Care answers questions for a fee but posts answers to commonly asked questions on his Web site. The information about bladder and kidney stones (calculi) in dogs and cats covers causes, signs and symptoms, treatment, and prevention.

Bladder Problems in Cats
<http://www.moggies.co.uk/articles/bladder.html>

"Moggies" is a U.K.-based site that organizes useful links and articles about cats and their care. The term *moggy* refers to a mixed breed or domestic cat. The information about bladder problems was written by Peter G. Knapp, a member of the Royal College of Veterinary Surgeons, and covers feline lower urinary tract disorder, formerly known as feline urological syndrome, diagnosis, and treatment.

Cats—Kidneys, Bladder, and Urinary System
<http://www.petplace.com/browse.aspx?a=2&c=94&parenttopic=CatMedical>

Veterinarian Jon Rappaport's site features separate pet centers for cats, dogs, birds, horses, reptiles, fish, and small mammals; drug library; medical center; veterinarian locator; and information about pet insurance. The information about kidney and urinary problems in cats covers cystitis, kidney failure, kidney stones, polycystic kidney disease, pyelonephritis, bladder cancer, and much more.

Feline Urological Syndrome
<http://maxshouse.com/feline_urological_syndrome_fus.htm>

Max's House celebrates the life of Max, a cat who passed away. Site features include a main subject index (ranging from life-threatening emergencies to understanding scratching); topic indexes (ranging from anatomy to virology); individual publications (ranging from spontaneous abortion to why cats have claws); and more.

Kidney Disease
<http://www.ah.novartis.com/pet_health/en/cat/cri/kidney_disease.shtml>

Novartis Animal Health manufactures a wide variety of products that benefit pet animals and farm animals, including treatments for heartworm, fleas, and roundworms. The section about kidney disease in cats covers late stage symptoms of chronic kidney diseases.

BONES, JOINTS, AND MUSCLES

Arthritis in the Feline
<http://www.isabellevets.co.uk/health_advice/cat/info/arthritiscat.htm>

IsabelleVets is a U.K.-based veterinary practice. Their site information about arthritis in cats covers degenerative and inflammatory forms of arthritis, and its diagnosis, treatment, and possible causes.

The Body of the Cat
<http://www.moggies.co.uk/stories/catbody.html>

"Moggies" is a U.K.-based site that organizes useful links and articles about cats and their care. The basic anatomical features of the adult cat are explained in text and line drawings.

DECLAWING—What You Need to Know
<http://www.declawing.com/htmls/declawing.htm>

Veterinarian Christianne Schelling explains declawing as a form of amputation, shows the anatomy of the claw, discusses the surgical procedures involved, and details possible postoperative complications.

Declawing Your Cat
<http://www.talktothevet.com/ARTICLES/CATS/catdeclawing.HTM>

TalkToTheVet.com is a site that features veterinarians' answers to commonsense problems experienced by pet owners. The section about cat declawing (also called onychectomy) details the surgical procedure, home care, and possible complications.

FAQs on Cat Declawing
<http://www.pawproject.com/html/faqs.asp>

The Paw Project is a nonprofit organization whose "mission is to educate the public about the painful and crippling effects of feline declawing, to promote animal welfare through the abolition of the practice of declaw surgery, and to rehabilitate big cats that have been declawed." The frequently asked questions (FAQs) related to this subject cover the reasons why cats scratch in the first place, the surgical procedure involved, potential complications, behavioral alternatives to declawing a cat, claw caps, and much more.

Musculoskeletal System of the Cat
<http://www.pawsonline.info/anatomy/musculoskeletal_system.htm>

Paws Online is a site that focuses on cat anatomy, physiology, behavior, and care. The information about the feline musculoskeletal system includes a clear drawing of muscles, bones, and joints.

Pet Arthritis Center
<http://www.glucosamine-arthritis.org/arthritis-pets/index.html>

The Arthritis & Glucosamine Information Center's Pet Arthritis Center includes a product guide that evaluates glucosamine formulations; hip dysplasia; types of arthritis that afflict pet animals; the susceptibility of large breed dogs to joint problems; and much more. The information specific to cats relates to hip dysplasia and signs of arthritis.

BRAIN AND NERVES

Seizures in Cats
<http://www.sniksnak.com/cathealth/seizures.html>

Pawprints and Purrs, Inc. is a Louisiana-based nonprofit organization that concerns itself with the welfare and well-being of cats and dogs. This information explains seizures, how they can be prevented or treated, and the importance of following a veterinarian's instructions.

Vestibular Disease in Cats
<http://www.petstyle.com/cat/health_well_article.aspx?id=365§ion=Health>

PetStyle's information about vestibular problems explains that the vestibular system governs coordination and balance. In cats, vestibular problems can be caused by cancer, inflammation, or infection, usually originating in the ear or brain. This content covers signs and symptoms (walking difficulties including the tendency to stumble or circle, and facial tics) of this disease, and recommended treatment.

CANCER

Cancer in Animals
<http://www.cancerindex.org/clinks30.htm>

Cancer Index organizes cancer-related Internet links into categories from cancer types, treatments, locations, and more. The links related to animals cover research, tumor registries, treatment, pet loss support groups, and more.

Cancer in Dogs and Cancer in Cats
<http://www.pets.ca/articles/article-cancer.htm>

Pets.ca is a Canadian site that provides pet information. Veterinarian Ron Hines explains how normal cells become cancerous, how biopsies are used to determine whether a tumor is cancerous or benign, common types of cancer in dogs and cats, and some treatment options.

Cancer in Pets
<http://www.southpaws.com/topics/cancer.html>

SouthPaws Veterinary Specialists & Emergency Center in Fairfax, Virginia offers this information about pet cancers, written by veterinarian Sarah E. Sheafor. Dr. Sheafor states that cancer treatment depends on type, stage, and location of the cancer, and briefly explains how diagnoses and staging workups are made.

Feline Cancer Overviews
<http://www.zzcat.com/resources/overview.htm>

This information clearinghouse was collected and organized by an individual whose cats have suffered from cancer. Site features include cancer signs, symptoms, and diagnosis; overview of cancers occurring in the mouth, gastrointestinal system, lymph glands, mammary glands, skin, vaccine injection sites, etc.; treatment options (surgery, chemotherapy, radiation); nutritional support; euthanasia decisions; and support for grief and loss.

Veterinary Cancer Society
<http://www.vetcancersociety.org/>

Veterinarians who specialized in cancer care formed this organization in 1974. The organization's site focuses on clinical trials for animals, finding specialists in specific geographic areas (United States and international), and links to external sites.

What Is Cancer in Cats?
<http://www.petplace.com/cats/what-is-cancer-in-cats/page1.aspx>

Veterinarian Jon Rappaport's information about cancer in cats covers basal cell skin cancer, chondrosarcoma of the nasal and paranasal sinus, chondrosarcoma of the larynx and trachea, fibrosarcoma, hemangiosarcoma, injection-site sarcoma, melanoma, leukemia, lipoma, mammary gland tumors, ovarian tumors, kidney tumors, lung tumors, and the like.

What You Should Know About Cancer in Animals
<http://www.avma.org/communications/brochures/cancer/cancer_brochure.asp>

The American Veterinary Medical Association (AVMA) worked with the Veterinary Cancer Society to develop this brochure. It is available in English and Spanish languages and explains diagnosis of cancer, common types (skin, breast, head and neck, lymphoma, feline leukemia complex, testicles, abdominal tumors, and bone tumors), and ten common cancer signs in pet animals.

DIABETES MELLITUS

Diabetes Mellitus
<http://www.marvistavet.com/html/diabetes_mellitus.html>

The Mar Vista Animal Medical Center in Los Angeles, California offers a wide range of pet health information on their site. The information about diabetes mellitus includes a disease overview, explanation of how insulin is administered, and insulin alternatives.

Diabetes Mellitus in Cats
<http://www.peteducation.com/article.cfm?cls=1&cat=1328&articleid=196>

Veterinarians Foster and Smith's site explains diabetes mellitus in cats in terms of causes and the likelihood of the disease occurring in older, overweight male cats.

Feline Diabetes
<http://www.vet.cornell.edu/fhc/brochures/diabetes.html>

Cornell University College of Veterinary Medicine created this comprehensible brochure about diabetes mellitus in cats, including how the disease is diagnosed, treating diabetes at home, monitoring blood glucose levels, and prognostic factors. This section also links to a video clip about caring for a diabetic cat.

Feline Diabetes—Diabetes in Cats
<http://www.felinediabetes.com/>

Established in 1996 by medical doctor Rebecca Price, features of this site include frequently asked questions, health articles, a Web-based discussion board, Rebecca's Diary, Sugarcat Journals, and poetry. The site also includes a dictionary of terms, blood glucose converter calculator, insulin logs that can be printed and used, information sheets for pet sitters that can be printed and used, and more. The cat health articles related to diabetes include an overview of the disease, treatment options, different types of insulin and their administration, blood glucose testing, recognizing low blood sugar, and the importance of diet when controlling diabetes.

EARS, EYES, AND NOSE

Ear Problems in the Cat
<http://www.fabcats.org/earproblems.html>

Feline Advisory Board (FAB) is a U.K.-based charitable organization that is "dedicated to promoting the health and welfare of cats through improved feline knowledge." The information about ear problems shows line drawings of the external, middle, and inner ear; and explains the symptoms and treatment of common ear diseases such as wounds, blood blisters, mange, mites, and more.

Eye Problems and Disease in Cats
<http://www.2ndchance.info/cateye.htm>

Veterinarian Ron Hines offers understandable information about eye diseases in cats including feline herpes virus, chlamydia, bartonella, and mycoplasma.

Feline Uveitis
<http://www.animaleyecare.net/diseases/uveitis.htm>

Animal Eye Care LLC is located in Bellingham, Washington. Its site includes information about uveitis in cats. Uveitis is an inflammation of the pigmented part of the cat's eye. The site explains diagnosis, causes, and treatment.

Progressive Retinal Atrophy/Degeneration
<http://www.peteducation.com/article.cfm?cls=1&cat=1314&articleid=342>

Veterinarians Foster and Smith's site information about retinal atrophy/degeneration in dogs and some cats explains and illustrates the anatomy of the eye (cornea, conjunctiva, iris, lens, vitreous, retina, and optic nerve), the signs of retinal atrophy or degeneration, and the progressive blindness that results from this disease.

HEART, ARTERIES, AND LUNGS

Cardiomyopathy in Cats
<http://www.sniksnak.com/cathealth/cardio.html>

Pawprints and Purrs, Inc. is a Louisiana-based nonprofit organization that concerns itself with the welfare and well-being of cats and dogs. The information about heart muscle disease (cardiomyopathy) in cats was written by a veterinarian and covers signs and symptoms, diagnosis, treatment, possible complications, and long-term outcomes.

The Coughing Pet
<http://www.vetmed.wsu.edu/ClientED/cough.asp>

Washington State University's College of Veterinary Medicine site includes pet health topics (bunny care, adopting a pet, determining the sex of a kitten, etc.), anatomy (text and images related to dog or cat organ systems), and definitions and abbreviations of common veterinary terms. Coughing is more frequent in dogs than cats. Possible causes and treatments are detailed in this site.

Heart and Blood Vessels
<http://www.petplace.com/browse.apx?a=2&c=92&parenttopic=CatMedical>

Veterinarian Jon Rappaport's site information about the cardiovascular system discusses the imprecision of the term heart attack when describing sudden death, fainting, syncope, or seizures. This section describes diagnosis, treatment, and home care for pets suffering from heart rhythm, low blood pressure, and other cardiac problems.

INFECTIONS

Avian Influenza (Bird Flu)
<http://www.cdc.gov/flu/avian>

The U.S. Centers for Disease Control and Prevention (CDC) is concerned with the monitoring and prevention of disease. CDC information about influenza in birds covers key facts (statistics, symptoms, complications, transmission, vaccinations). Infected migratory birds may pose a threat to pet animals.

Bird Flu—A Danger to Felines?
<http://www.aafponline.org/WhatsNew/bird_flu.htm>

American Association of Feline Practitioners (AAFP) is a professional organization of veterinarians who practice feline medicine and surgery. Veterinarian James Richards of Cornell Feline Health Center explains whether bird flu poses any risks to predatory cats or through uncooked poultry products.

Ear Infections in Cats
<http://www.petstyle.com/cat/health_well_article.aspx?id=350>

PetStyle provides useful information about feline ear infections caused by bacteria, fungi, or yeasts.

Feline Abscesses
<http://www.vetcentric.com/reference/encycEntry.cfm ?ENTRY=65&COLLECTION=EncycIllness&MODE=full>

VetCentric's information about bite-related abscesses covers signs, diagnosis, transmission, prognosis, treatment, and prevention.

Feline Immunodeficiency Virus, FIV
<http://www.vetcentric.com/reference/encycEntry.cfm ?ENTRY=62&COLLECTION=EncycIllness&MODE=full>

VetCentric's information about feline immunodeficiency virus includes a disease overview, signs, diagnosis, prognosis, transmission, treatment, and prevention.

Feline Infectious Anemia
<http://www.marvistavet.com/html/body_feline_infectious _anemia.html>

The Mar Vista Animal Medical Center in Los Angeles, California offers a wide range of pet health information on their site. Feline infectious anemia, also called feline hemotropic mycoplasmosis, is caused by *Hemobartonella felis* or *Mycoplasma haemofelis*. Site information covers diagnostic procedures, risks to outdoor cats, and treatment.

Feline Infectious Peritonitis (FIP)
<http://www.peteducation.com/article.cfm?cls=1&cat=1316&articleid=212>

Veterinarians Foster and Smith's information about feline infectious peritonitis (FIP) covers causes, transmission, signs and symptoms, testing, diagnosis, treatment, and prevention.

Feline Infectious Peritonitis and Pleuritis
<http://www.merckvetmanual.com/mvm/index.jsp?cfile=htm/bc/56900.htm>

Drug manufacturer Merck & Co., Inc.'s veterinary manual includes more than 12,000 topics including feline infectious peritonitis. This section details the causes (strains of the coronavirus), transmission, signs, diagnosis, and prevention of feline infectious peritonitis, also known as feline coronaviral vasculitis.

Feline Infectious Peritonitis, FIP
<http://www.vetcentric.com/reference/encycEntry.cfm?ENTRY=20&COLLECTION=EncycIllness>

VetCentric's information about feline infectious peritonitis explains how it affects both wild and domestic felines, and gives an overview, signs, symptoms, diagnosis, prognosis, transmission, treatment, and prevention.

Feline Panleukopenia, Feline Parvovirus, Feline Distemper
<http://www.vetcentric.com/reference/encycEntry.cfm?ENTRY=19&COLLECTION=EncycIllness&MODE=full>

VetCentric's information about panleukopenia (distemper) in cats includes an overview, signs, symptoms, diagnosis, prognosis, transmission, treatment, and prevention. Feline distemper is caused by the parvovirus.

H5N1 in Cats
<http://www.fao.org/ag/againfo/subjects/en/health/diseases-cards/avian_cats.html>

The Animal Production and Health Division of the Food and Agriculture Organization of the United Nations developed this information to explain the possible transmission of H5N1 (influenza A) virus from an infected bird to a domestic or wild cat, and the public health implications of such transmissions.

Infectious Upper Respiratory Complex
<http://www.vetcentric.com/reference/encycEntry.cfm?ENTRY=24&COLLECTION=EncycIllness&MODE=full>

VetCentric's information about upper respiratory infections in cats explains that these infections are caused by feline herpesvirus (sometimes called feline viral rhinotracheitis), feline calcivirus, and occasionally by *Chlamydia psittacci,* which also causes conjunctivitis. Site information includes an overview, signs, symptoms, diagnosis, prognosis, transmission, treatment, and prevention.

Panleukopenia (Feline Distemper)
<http://www.peteducation.com/article.cfm?cls=1&cat=1314&articleid=222>

Veterinarians Foster and Smith's information about panleukopenia (feline distemper) includes transmission, symptoms, diagnosis, treatment, and prevention.

Pet Pests—Infections and Pets
<http://us.pampers.com/en_US/content/type/101/contentId/9269.do>

The manufacturers of Pampers disposable diapers developed this useful information about the risks that cats, dogs, turtles, birds, and hamsters pose to humans, especially children and pregnant women. Physician Angela Rosas explains the risks of fleas, ticks, ringworm, animal bites, bacteria, toxoplasmosis, and salmonella.

Rabies
<http://www.nlm.nih.gov/medlineplus/rabies.html>

MedlinePlus is a consumer health resource developed by the U.S. National Library of Medicine that provides extensive information about more than 650 diseases and conditions. The information about rabies covers overviews, prevention/screening, organizations, statistics, and children.

Rabies in Cats
<http://www.peteducation.com/article.cfm?cls=1&cat=1316&articleid=346>

Veterinarians Foster and Smith's information about rabies in cats covers facts, risks, disease transmission, symptoms according to disease phase, diagnosis, treatment, and prevention.

Toxoplasmosis
<http://www.vetcentric.com/reference/encycEntry.cfm?ENTRY=21&COLLECTION=EncycIllness&MODE=full>

VetCentric's information about toxoplasmosis in cats covers signs of infection, symptoms, diagnosis, prognosis, transmission, treatment, prevention, and risks to pregnant or immunocompromised individuals.

Toxoplasmosis
<http://www.peteducation.com/article.cfm?cls=1&cat=1314&articleid=770>

Veterinarians Foster and Smith's information about toxoplasmosis covers disease transmission and signs and symptoms in cats, dogs, and humans.

West Nile Virus Information
<http://www.nal.usda.gov/awic/news/wnvirus/wnfacts.htm>

Animal Welfare Information Center (AWIC) at the National Agricultural Library of the U.S. Department of Agriculture released these facts about West Nile virus including its transmission and protecting oneself and pet animals. Updated site content can be delivered by RSS feed.

West Nile Virus
<http://www.animalhealthcare.ca/contents/content_careguides2.asp?get=faq&id=124&category=40>

The Canadian Veterinary Medical Association developed this site with pet health content in English and French languages. Site content related to West Nile virus covers specific animals that can become infected, how the infection is diagnosed, and relative risks.

OLD AGE

☑ **Feline Senior Care Guidelines**
<http://www.aafponline.org/resources/guidelines/senior/FelineSeniorCareGuidelines_27pages.pdf>

American Association of Feline Practitioners (AAFP) is a professional organization of veterinarians who practice feline medicine and surgery. This twenty-seven-page document outlines how veterinarians should care for senior cats, specific diseases of concern in older cats (hyperthyroidism, chronic renal failure, hypertension, cancer, diabetes mellitus, and inflammatory bowel disease), behavioral problems, pain management, use of anesthetic agents in older cats, diet and nutrition especially in cats with specific diseases, pet loss, and much more (Figure 4.1).

Geriatric Cats—Special Needs
<http://www.sniksnak.com/cathealth/geriatric.html>

Pawprints and Purrs, Inc.'s information about caring for older cats stresses the importance of potassium balance, and signs of aging and disease (weight loss, vomiting, changes in elimination, etc.).

Keeping Your Senior Cat Healthy
<http://www.petplace.com/cats/keeping-your-senior-cat-healthy/page1.aspx>

Veterinarian Jon Rappaport's information about keeping older cats healthy covers the importance of keeping an older cat groomed, active, well nourished, and well rested; and scheduling regular veterinarian visits.

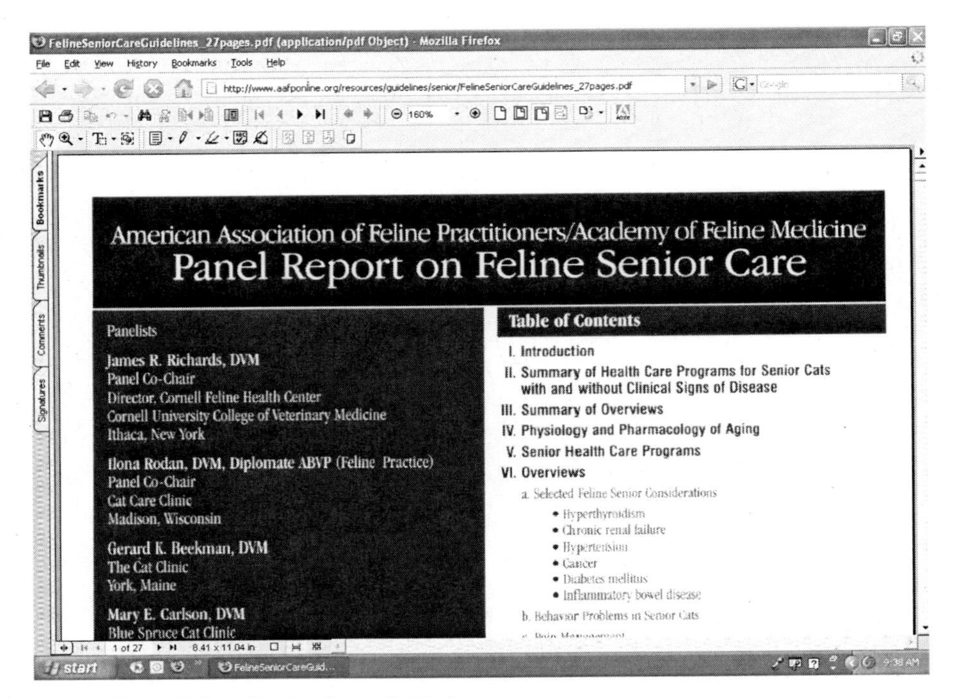

FIGURE 4.1. Feline Senior Care Guidelines <http://www.aafponline.org/resources/guidelines/senior/table_of_contents.pdf> Reprinted with permission.

Senior Pet Care <http://www.healthypet.com/library_view.aspx?id=141>

Healthypet is a site that was developed and is maintained by the American Animal Hospital Association (AAHA). Care tips for senior pets include the value of senior health examinations and baseline laboratory tests (complete blood count, urinalysis, blood-chemistry panel, and parasite evaluation), and diets formulated for older pets. Effects of aging include slower response times and physical changes (inappropriate elimination or incontinence).

The Special Needs of the Senior Cat <http://www.vet.cornell.edu/fhc/brochures/seniorcat.html>

Cornell University's College of Veterinary Medicine developed this brochure to explain that although aging is a process rather than a disease, older cats experience a host of physical and behavioral changes. Tips for keeping senior cats healthy and happy include daily grooming, proper nutrition, adequate exercise, and a relatively stress-free environment.

PARASITES

Parasites that can cause diseases in cats include worms, mites, ticks, fleas, and other organisms. If left untreated, these pests can cause complications and in some cases, spread to humans.

Parasites in Cats
<http://www.peteducation.com/category_summary.cfm?cls=1&cat=1357>

Veterinarians Foster and Smith describe diseases caused by fleas, ticks, mites, heartworms, bladder worms, eyeworms, giant kidney worms, trichinosis, coccidia, giardia, tapeworms, roundworms, whipworms, and more. Fire ants, lice, and types of flies can also affect cats. This section also explains fecal examinations, deworming schedules, tick and flea control, heartworm prevention, and transmission of roundworms and hookworms to people.

Feline Heartworm Disease
<http://www.vetcentric.com/reference/encycEntry.cfm?ENTRY=44&COLLECTION=EncycIllness&MODE=full>

VetCentric's section related to heartworm disease in cats covers its transmission by mosquitoes; signs and symptoms of the disease; diagnosis; prognosis; treatment; and prevention.

Fleas, Ticks, and Mites
<http://www.animalhealthchannel.com/fleasticksmites/>

Animalhealthchannel "provides comprehensive, trustworthy information about conditions and diseases that affect animals, as well as pet care and training information," and its content is "developed and monitored by board-certified veterinarians." Fleas, ticks, and mites are described as parasites that can cause allergic reactions and diseases such as Lyme disease, mange, and hair loss.

Heartworm
<http://www.marvistavet.com/html/heartworm.html>

The Mar Vista Animal Medical Center in Los Angeles, California offers a wide range of pet health information on their site. The section on heartworm explains the *Dirofilaria immitis* parasite, the distinction between heartworm infection and heartworm disease, diagnosis, treatment, and prevention for dogs and cats.

Hookworms
<http://www.marvistavet.com/html/hookworms.html>

The Mar Vista Animal Medical Center site's section on hookworms explains the *Ancylostoma caninum* or *Ancylostoma braziliense* parasites, lifecycle of the parasite, how these worms get transmitted from dog mother to puppies, and how humans can be infected.

Parasitic Worms
<http://www.ah.novartis.com/pet_health/en/cat/worms/index.shtml>

Novartis Animal Health manufactures a wide variety of products that benefit pet animals and farm animals, including treatments for heartworm, fleas, and roundworms. Site information focuses on detecting, treating, and preventing fleas, parasitic worms, separation anxiety, atopic dermatitis, heart diseases, and kidney diseases.

Tapeworms
<http://www.vetcentric.com/reference/encycEntry.cfm?ENTRY=64&COLLECTION=EncycIllness&MODE=full>

VetCentric's section about tapeworms (cestodes) provides an overview of the organisms involved (*Dipylidium caninum, Taenia taeniaformis, Taenia pisiformis, Taenia hydratigena,* and *Echinococcus multilocularis*), signs and symptoms, diagnosis, prognosis, treatment, and prevention.

Worms
<http://www.animalhealthchannel.com/worms/>

Animalhealthchannel "provides comprehensive, trustworthy information about conditions and diseases that affect animals, as well as pet care and training information," and its content is "developed and monitored by board-certified veterinarians." Site content briefly describes roundworms, whipworms, tapeworms, hookworms, and heartworms; and transmission, symptoms, diagnosis, and treatment.

SKIN, FUR, AND COAT

Feline Miliary Dermatitis
<http://www.sniksnak.com/cathealth/mil-derm.html>

Pawprints and Purrs, Inc.'s site information about miliary dermatitis explains common symptoms, diagnosis, and treatment. This disease manifests itself as crusty lesions on a cat's skin and can be caused by parasites, bacteria, fungi, yeasts, autoimmune diseases, allergic reactions, and much more.

Feline Scabies
<http://www.peteducation.com/article.cfm?cls=1&cat=1334&articleid=737>

Veterinarians Foster and Smith explain that feline scabies (notoedric mange) is caused by the *Notoedris cati* mite, the parasite's life cycle, transmission (cat-to-cat), symptoms, diagnosis, treatment, and prevention.

Mange
<http://edis.ifas.ufl.edu/BODY_MG118>

University of Florida's Institute of Food and Agricultural Sciences developed this brief information to explain mange mites in dogs and cats, including sarcoptic mange, demodectic mange (or red mange), and ear mange.

Mange in Dogs and Cats
<http://www.merckvetmanual.com/mvm/index.jsp?cfile=htm/bc/72005.htm>

Drug manufacturer Merck & Co., Inc.'s veterinary manual includes more than 12,000 topics including sarcoptic mange in dogs, notoedric mange in cats, otodectic mange in the ears of cats or dogs, cheyletiellosis (walking dandruff), which infects cats, dogs, or rabbits; demodicosis, which afflicts dogs or cats; and trombiculosis, found in cats and dogs. Each type of mange is explained in terms of the parasites involved, diagnosis, and management.

STOMACH, INTESTINES, AND DIGESTIVE SYSTEM

Anal Sacs
<http://www.marvistavet.com/html/anal_sacs.html>

The Mar Vista Animal Medical Center site uses clever animations to show the location and procedure for emptying the glands around a dog's anus. Site content explains the significance of "scooting" and the importance of expressing material from these glands from either cats or dogs.

Digestive System
<http://www.peteducation.com/category_summary.cfm?cls=1&cat=1322>

Veterinarians Foster and Smith's information about the cat digestive system covers anatomy of the esophagus, stomach, and intestines; common intestinal parasites; diaphragmatic hernias; foreign bodies; hairballs; intestinal gas; rectal prolapse; and related topics.

Feline Pancreatitis
<http://www.sniksnak.com/cathealth/pancreatitis.html>

Pawprints and Purrs, Inc.'s information about inflammation of the pancreas in cats covers the role that the pancreas plays in digestion; causes of inflammation, signs, treatment, and long-term outcomes.

Inflammatory Bowel Disease
<http://www.marvistavet.com/html/body_inflammatory_bowel_disease.html>

The Mar Vista Animal Medical Center's site content related to inflammation of the colon, resulting in severe vomiting and diarrhea includes diagnosis, causes, and treatment.

Stomach, Intestine, Liver, or Pancreas
<http://www.petplace.com/browse.aspx?a=2&c=116&parentTopic=CatMedical>

Veterinarian Jon Rappaport's site information about the cat digestive system covers pancreatitis, gastrointestinal diseases related to diet, foreign objects, inflammation of the esophagus, hairballs, liver diseases, inflammatory bowel disease, pica, rectal prolapse, and more.

TEETH AND GUMS (See Also DENTAL CARE)

Cat Dental Care
<http://www.caws.org/dentalcat.html>

Community Animal Welfare Society (CAWS) based in Salt Lake City, Utah is a nonprofit group that is dedicated to animal welfare and rescue. Site information related to dental care for cats includes teeth cleaning and tooth brushing.

Cat Exam
<http://www.pet-care-plus.com/cat-care/cat-health/cat-exam.htm>

Pet Care Plus is a site that focuses on the care and feeding of dogs, cats, birds, and exotic pets (rabbits, chinchillas, tropical fish, guinea pigs, hamsters, rats, prairie dogs, hedgehogs, frogs, turtles, snakes, lizards, tarantulas). The section devoted to cat care includes information about at-home physical examinations.

Dental Care Series
<http://www.veterinarypartner.com/Content.plx?P=SRC&S=1&SourceID=13>

VeterinaryPartner.com is a site that has been developed and maintained by veterinarians and veterinary experts associated with the Veterinary Information Network (VIN). The information this dental care series covers tooth brushing, ceramic materials to restore tooth enamel, periodontal disease, orthodontics, and much more. The information about feline oral resorptive lesion (FORL), stomatitis, and oral cancers was written by veterinarian Jan Bellows of the All Pets Dental Clinic in Pembroke Pines, Florida.

Feline Dental Disease
<http://www.sniksnak.com/cathealth/dental2.html>

Pawprints and Purrs, Inc.'s feline dental disease section covers tooth tartar, teeth cleaning, and the importance of diet.

Oral and Dental Anatomy of Dogs, Cats, and Ferrets
<http://www.peteducation.com/article.cfm?cls=1&cat=1369&articleid=325>

Veterinarians Foster and Smith's information about oral and dental anatomy explains the tongue and gums; salivary glands; the upper and lower jaws; types of teeth common to meat-eating mammals (incisors, canines, premolars, and molars); numbers of teeth (deciduous and permanent) for dogs, cats, and ferrets; and parts of the tooth.

Chapter 5

For Dogs Only

My little dog—a heartbeat at my feet.

Edith Wharton

GENERAL DOG HEALTH SITES

AKC Canine Health Foundation Page
<http://www.akcchf.org/>

The motto of the American Kennel Club (AKC) Canine Health Foundation (CHF) is "working together for the health of your dog." The AKC CHF site features information about health help (diseases and conditions listed by disease or breed), news and events, research (articles and white papers), and more.

Animal Planet—Dog Guide
<http://animal.discovery.com/guides/dogs/dogs.html>

Animal Planet features a guide for dog owners that covers ears and hearing, smell and taste, hair of the dog, paws, skulls, teeth, staying cool, life stages, choosing mixed breeds versus purebreds, feeding, exercise, grooming, pregnancy, aging, veterinary care, behavior, training, safety, and much more.

Canine Health Information Center (CHIC)
<http://www.caninehealthinfo.org/>

CHIC is a joint effort of the American Kennel Club's Canine Health Foundation and the Orthopedic Foundations for animals. Site features

Internet Guide to Pet Health

doi:10.1300/5675_05

include a searchable database of pure breeds and associated conditions. For example, the database record for Alaskan Malamute links to information about hip dysplasia, eye clearance, and autoimmune thyroiditis. Although this information is intended for breeding research purposes as CHIC maintains a DNA bank, diseases and conditions associated with certain breeds will be of interest to pet owners (Figures 5.1 and 5.2).

Dog Owner's Guide <http://www.canismajor.com/dog/>

Dog Owner's Guide includes "more than 300 pages of features, breed profiles, training tips, health information, and articles about shelters, rescue, dogs, and the law." Link highlights include choosing a dog, caring for a new puppy, breed profiles, canine aggression, working dogs, and more.

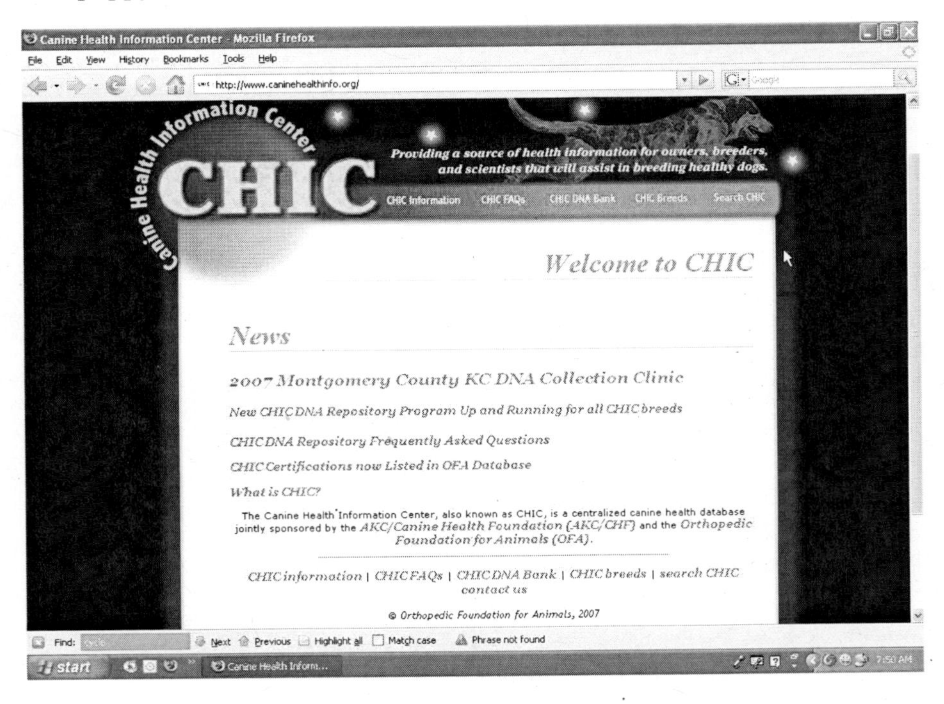

FIGURE 5.1. Canine Health Information Center (CHIC) <http://www.caninehealthinfo.org>
Reprinted with permission of www.offa.org and www.caninehealthinfo.org.

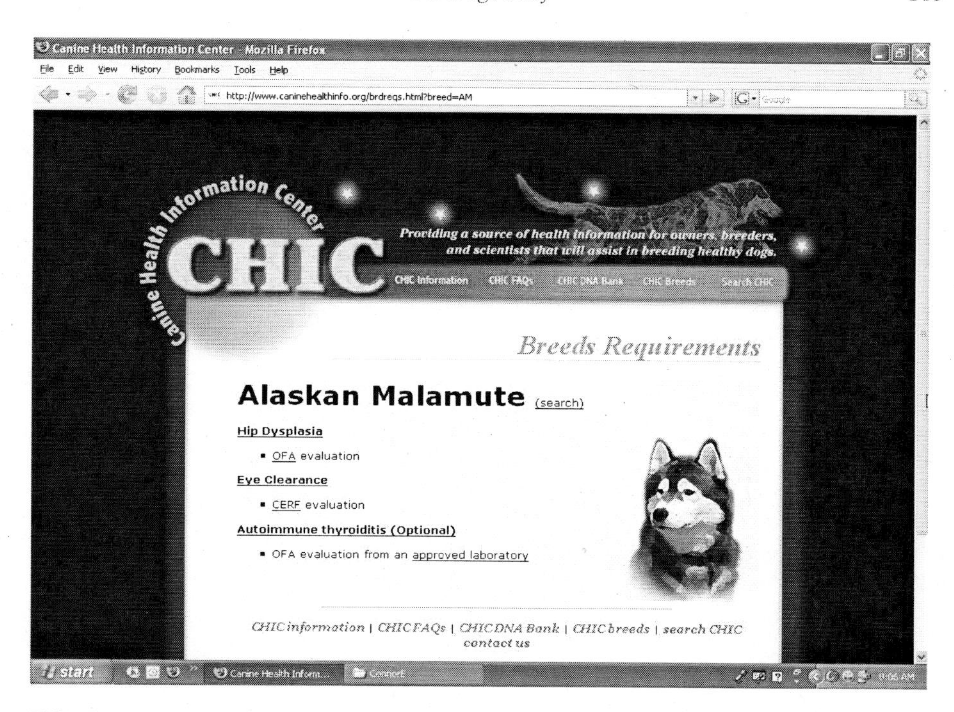

FIGURE 5.2. Canine Health Information Center (CHIC)—Alaskan Malamute <http://www.caninehealthinfo.org/brdreqs.html?breed=AM>
Reprinted with permission of www.offa.org and www.caninehealthinfo.org.

☑ DogAge Canine Care—Fast Facts <http://www.dogage.com/care/factsIntro.aspx>

The developers of RealAge, a resource that helps humans calculate their "real" physiological age based on a number of lifestyle factors, developed this site to help humans calculate the "real" ages of their pet dogs by applying this same principle to dogs. The DogAge site also includes helpful information about caring for dogs: preventive care (weight control, vaccinations, dental care, etc.); various diseases and conditions (eye problems, hearing problems, joint pain, separation anxiety, cognitive dysfunction syndrome, cancer, epilepsy, heartworm, intestinal worms, respiratory infections, skin infections); diet, exercise, and safety.

SPECIFIC DISEASES AND CONDITIONS IN DOGS

Dogs can suffer from a variety of diseases and conditions, some of which are chronic and life threatening. Some breeds are predisposed to certain behavioral and physical maladies and conditions, and other health issues related to age.

BEHAVIORAL PROBLEMS

Some of the leading causes for animal abandonment are behavioral such as excessive barking, inappropriate elimination, and/or aggression toward humans. Exercise, training, and play activities can be used to modify behavior in pet animals.

Canine Separation Anxiety
<http://www.cpvh.com/Articles/47.html>

Claws & Paws Veterinary Hospital in Pearland, Texas, features useful information of interest to pet owners including an explanation of why dogs get anxious when separated from their owners, and ways to remedy these problems (whining, inappropriate elimination, chewing, etc.) through medication and/or behavior modification.

Guide to Behavior Problems in Puppies
<http://petplace.netscape.com/articles/artShow.asp?artID=2990>

Veterinarian Jon Rappaport's information about behavior problems in puppies covers inappropriate urination and/or defecation, urine marking, submissive urination, digging, chewing, separation anxiety, fears (people, other dogs), aggressive behaviors (nipping, biting, growling, lunging), barking, running away, and more.

Pet Behavior Problems—Inappropriate Urination
<http://www.vet4petz.com/articles/inappropriate_urination.htm>

Vet4Petz is a site that offers useful pet health information for site members and nonmembers alike. The information about pet behavior includes inappropriate urination/defecation by cats and dogs.

Rescue Dog Behavior Problems
<http://www.naiaonline.org/body/articles/archives/resc95gc.htm>

Veterinarian Gary L. Clemons explains behavior problems that may surface in dogs that have been rescued and medical conditions (hip dysplasia, arthritis conditions, parasites, malnutrition) that can trigger behavior problems.

Separation Anxiety
<http://www.ygrr.org/doginfo/behavior-anxiety.html>

Yankee Golden Retriever Rescue is a nonprofit organization based in Hudson, Massachusetts, that focuses on "providing veterinary care, comprehensive evaluation, and adoptive homes to Golden Retrievers who are abandoned, mistreated, neglected, or left in pounds throughout the six New England states (Connecticut, Maine, Massachusetts, New Hampshire, Rhode Island, & Vermont)." Their site includes useful information about dog safety, health, training, behavior, care, and death. The section related to dog behavior explains why dogs become anxious when separated from their owners.

BLADDER, KIDNEYS, AND URINARY TRACT

Bladder and Kidney Stones in Dogs and Cats
<http://www.2ndchance.info/calculi.htm>

Veterinarian Ron Hines of All Creatures Care answers questions for a fee but posts answers to commonly asked questions on his Web site. The information about bladder and kidney stones (calculi) in dogs and cats covers causes, signs and symptoms, treatment and prevention.

Bladder Stones (Uroliths)
<http://www.sniksnak.com/doghealth/bladderstones.html>

Pawprints and Purrs, Inc. is a Louisiana-based nonprofit organization that concerns itself with the welfare and well-being of cats and dogs. This information explains how bladder stones (uroliths) in dogs are formed, diagnosed, treated, and prevented.

Canine Prostatic Disease
<http://www.sniksnak.com/doghealth/prostate.html>

Pawprints and Purrs, Inc.'s information about prostate diseases in dogs explains the signs and symptoms, disease process, diagnosis, and treatment.

Urinary Incontinence in Dogs
<http://www.peteducation.com/article.cfm?cls=2&cat=1634&articleid=487>

Veterinarians Foster and Smith's information about urinary incontinence in dogs covers causes, risks, and treatments.

BONES, JOINTS, AND MUSCLES

Arthritis/Joint Disease Center
<http://www.marvistavet.com/html/body_arthritis_joint_disease_center.html>

The Mar Vista Animal Medical Center in Los Angeles, California offers a wide range of pet health information on their site. The content related to arthritis and joint diseases covers the anatomy of normal joints, physical therapy for dogs with arthritic conditions, medications prescribed to treat arthritic conditions, and specific joint diseases (hip dysplasia, elbow dysplasia, osteochondrosis, and ruptured anterior cruciate ligament).

Cruciate Ligament Rupture
<http://www.sniksnak.com/doghealth/cru-ligament.html>

Pawprints and Purrs, Inc.'s information about knee injuries in dogs explains causes of the injury, diagnosis, and treatment.

Orthopedic Foundation for Animals (OFA)
<http://www.offa.org/>

OFA collects information about orthopedic diseases in dogs, and is "dedicated to the improved health and well-being of companion animals."

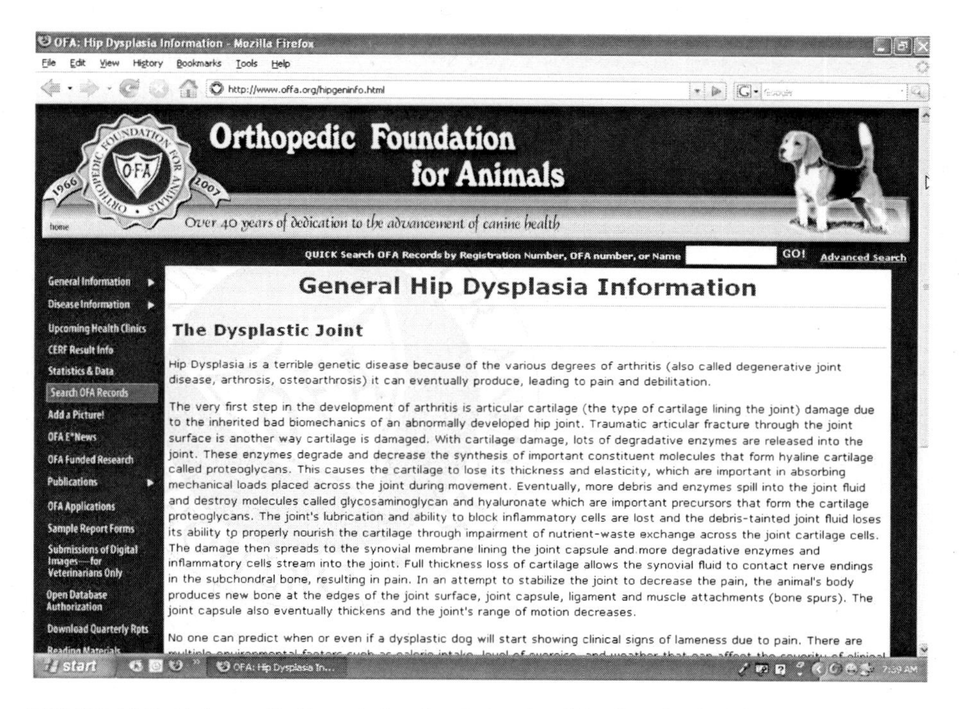

FIGURE 5.3. Orthopedic Foundation for Animals—Hip Dysplasia Information <http://www.offa.org/hipgeninfo.html>
Reprinted with permission.

Site features include breaking news, general information about the foundation, disease information (hip dysplasia, elbow dysplasia, kneecap dislocation, Legge-Calve-Perthes disease of the hip joint, and other conditions), statistics by dog breed or disease, publications, and more (Figure 5.3).

The Pet Arthritis Center <http://www.glucosamine-arthritis.org/arthritis-pets/index.html>

The Arthritis & Glucosamine Information Center's Pet Arthritis Center includes a product guide that evaluates glucosamine formulations, hip dysplasia in dogs, hip dysplasia in cats, types of arthritis that afflict pet animals, the susceptibility of large breed dogs to joint problems, and much more.

BRAIN AND NERVES

Cognitive Dysfunction—Senility
<http://www.animalhospitals-usa.com/dogs/cognitive_dysfunction_senility.html>

Animal Hospitals USA's site features useful information about illnesses and injuries in "dogs, cats, birds, guinea pigs, gerbils, ferrets, prairie dogs, and reptiles." The information about senility in dogs covers cognitive dysfunction syndrome (or canine cognitive disorder), its symptoms (confusion, disorientation, changes in sleep patterns, etc.), and treatment.

Dementia (Senility, Cognitive Dysfunction)
<http://www.petplace.com/dogs/dementia-senility-cognitive-dysfunction/page1.aspx>

Veterinarian John McDonnell explains diagnosis, treatment, and home care for dogs with decreased cognitive functioning and signs of senility (slow response time, diminished interactions with family members, irritability).

Seizures in Dogs
<http://www.sniksnak.com/doghealth/seizures.html>

Pawprints and Purrs, Inc.'s information about seizure activity in dogs details the causes, phases, and prevention.

CANCER

Cancer in Animals
<http://www.cancerindex.org/clinks30.htm>

Cancer Index organizes cancer-related Internet links into categories from cancer types, treatments, locations, and more. The links related to cancer in animals cover research, tumor registries, treatment, pet loss support groups, and more.

Cancer in Cats and Dogs
<http://www.pets.ca/articles/article-cancer.htm>

Pets.ca is a Canadian site that provides pet information. The information about cancer in cats and dogs is written by veterinarian Ron Hines and explains how normal cells become cancerous, how biopsies are used to determine whether a tumor is cancerous or benign, common types of cancer in dogs and cats, and some treatment options.

Cancer in the Canine
<http://www.labbies.com/cancerintro.htm>

Wing-N-Wave Labradors is a breeder of Labrador retrievers. Their site information about cancer in dogs was written by Pamela A. Davol and is well organized and thorough. The material covers how normal cells become cancerous, various causes of cancer, detecting cancer in dogs, common cancers (skin, mammary glands, lymphoma, mouth tumors, and bone cancer), treatments and nutrition issues.

Veterinary Cancer Society
<http://www.vetcancersociety.org/>

Veterinarians who specialized in cancer care formed this organization in 1974. The organization's site focuses on clinical trials for animals and finding specialists in specific geographic areas (United States and international).

What You Should Know About Cancer in Animals
<http://www.avma.org/communications/brochures/cancer/cancer_brochure.asp>

The American Veterinary Medical Association (AVMA) worked with the Veterinary Cancer Society to develop this brochure in English and Spanish languages to explain diagnosis, common types (skin, breast, head and neck, lymphoma, feline leukemia complex, testicles, abdominal tumors, and bone tumors), and ten common cancer signs in pet animals.

DIABETES MELLITUS

Diabetes and Glucose Control
<http://www.peteducation.com/category_summary.cfm?cls=2&cat=1579>

Veterinarians Foster and Smith's site explains how glucose control in dogs can be achieved by frequent monitoring of glucose levels and administration of insulin.

Diabetes Mellitus
<http://www.merckvetmanual.com/mvm/index.jsp?cfile=htm/bc/40302.htm>

Drug manufacturer Merck & Co., Inc.'s veterinary manual includes more than 12,000 topics including this description of the causes, signs, diagnosis, and treatment of diabetes mellitus in animals.

Diabetes Mellitus
<http://www.sniksnak.com/doghealth/diabetes-mellitus.html>

Pawprints and Purrs, Inc.'s information about diabetes mellitus in dogs covers diabetes inspidus and diabetes mellitus, diagnosis, importance of insulin, treatment, and home monitoring.

EARS, EYES, AND NOSE

Cataracts
<http://www.purinaone.com/dogcare_cond_atoz_article.asp?Seed=691&ArticleNumber=30018>

Purina ONE is a brand of pet food. The information on this site about dogs covers how cataracts form, age of onset (at birth, early in life, later in life), breeds that are susceptible to cataracts, trauma, and treatment.

The Cold, Wet Nose
<http://www.northbay-canine.org/coldwet-txt.htm>

North Bay Canine Rescue & Placement in Petaluma, California offers light-hearted information about why a cold wet nose is not the most reliable indicator of good health in a dog, plants and substances that are poisonous to dogs, tips for keeping dogs comfortable during hot weather, essential first aid supplies, and signs and symptoms of illness. This information is organized into broad categories of disorder (ear, eye, mouth and tooth, respiratory, nervous, skin and coat, external parasites, blood and heart, reproductive, bone, muscle, and joint, urinary, and digestive).

Ear Care and Disease—Otitis Externa
<http://www.purinaone.com/dogcare_cond_atoz_article.asp ?Seed=680&ArticleNumber=30007>

Purina's information about external ear disease in dogs covers signs, causes (allergies, parasites, bacteria, yeast, foreign bodies, etc.), diagnosis, treatment, and preventive measures (ear cleaning, trimming hair).

Eyes, Ears, and Snout Maintenance
<http://www.seefido.com/html/eyes_ears_and_snout _maintenance.htm>

This site sells collars, leashes, harnesses and other training tools. The information about maintaining a dog's eyes, ears, and snout covers basic grooming.

Glaucoma in Dogs—An Eye Emergency
<http://www.peteducation.com/article.cfm?cls=2&cat= 1606&articleid=439>

Veterinarians Foster and Smith's information about glaucoma in dogs explains that it is a serious disease requiring emergency care. The disease is defined and described in terms of types (primary, secondary) and treatment.

Progressive Retinal Atrophy/Degeneration
<http://www.peteducation.com/article.cfm?cls=1&cat=1314&articleid=342>

Veterinarians Foster and Smith's information about retinal atrophy/degeneration in dogs and some cats explains and illustrates the anatomy of the eye (cornea, conjunctiva, iris, lens, vitreous, retina, and optic nerve), the signs of retinal atrophy or degeneration, and the progressive blindness that results from atrophy.

Snow Nose or Winter Nose in Dogs
<http://www.lowchensaustralia.com/grooming/snownose.htm>

Löwchens of Australia is a site that focuses on the Löwchen breed (little lion dogs). The site's simple explanation for the loss of pigmentation in a dog's snout is lack of sunlight during the winter months or vitamin deficiency.

HEART, ARTERIES, AND LUNGS

Canine Dilated Cardiomyopathy and Heart Failure
<http://www.sniksnak.com/doghealth/heartfailure.html>

Pawprints and Purrs, Inc. is a Louisiana-based nonprofit organization that concerns itself with the welfare and well-being of cats and dogs. The information about heart muscle disease in dogs explains various signs of heart failure, diagnostic tests, and treatments.

Heart and Circulatory System Disorders
<http://www.peteducation.com/category_summary.cfm?cls=2&cat=1610>

Veterinarians Foster and Smith's section on cardiovascular diseases discusses diagnosis, symptoms, treatment, and prevention of heartworm, aortic stenosis, heart valve diseases, pulmonary stenosis, and ventricular septal defect.

High Blood Pressure
<http://www.marvistavet.com/html/high_blood_pressure.html>

The Mar Vista Animal Medical Center's information about high blood pressure mentions associated diseases such as chronic renal failure, hyperthyroidism, Cushing's disease, and diabetes mellitus. The content explains how blood pressure is measured in pet animals, and possible treatment regimens.

Infectious Tracheobronchitis
<http://www.peteducation.com/article.cfm?cls=2&cat=1556&articleid=452>

Veterinarians Foster and Smith provide information about tracheobronchitis, also known as kennel cough, which is caused most commonly by parainfluenza, *Bordetella bronchisepta,* or mycoplasma. Treatment of kennel cough includes bronchodilators or cough-suppressants, and in more complicated cases, antibiotics.

INFECTIONS

Avian Influenza (Bird Flu)
<http://www.cdc.gov/flu/avian/>

The U.S. Centers for Disease Control and Prevention (CDC) is concerned with the monitoring and prevention of disease. The information about influenza A (H5N1), which can be spread from birds to humans, includes key facts, the current situation of the outbreak, infection in humans, information for specific groups (health professionals, state health departments, poultry industry, travelers, Americans living abroad, etc.), and much more. Of particular interest to pet owners is the information about susceptibility of domestic cats to influenza A infection.

Distemper
<http://www.peteducation.com/article.cfm?cls=2&cat=1556&articleid=419>

Veterinarians Foster and Smith explain that distemper is caused by paramyxovirus, which is similar to the virus that causes measles in humans.

Distemper affects almost all organ systems in the body. Symptoms, disease risk, treatment, and prevention are mentioned in this site.

Hot Spots—Acute Moist Dermatitis
<http://www.peteducation.com/article.cfm?cls=2&cat=1592&articleid=447>

Veterinarians Foster and Smith explain the causes (allergies, mites, infections, etc.), treatment, and prevention of hot spots (moist dermatitis) in dogs. Technically, a hot spot is moist dermatitis caused by an allergic reaction to a flea or other parasite. The itching causes the dog to lick and scratch the areas afflicted, and thereby causes more inflammation. This section discusses causes, treatment, and prevention of this condition, which can plague dogs with dense undercoats.

Parvovirus
<http://www.peteducation.com/article.cfm?cls=2&cat=1556&articleid=467>

Veterinarians Foster and Smith provide information about parvovirus, a common but serious infection that causes bloody diarrhea and often, death within a few days in puppies. Information covers transmission, symptoms, diagnosis, treatment, and immunity/vaccination.

Pet Pests—Infections and Pets
<http://us.pampers.com/en_US/content/type/101/contentId/9269.do>

The manufacturers of Pampers disposable diapers developed this useful information about the risks that cats, dogs, turtles, birds, and hamsters pose to humans, especially children and pregnant women. Physician Angela Rosas explains the risks of fleas, ticks, ringworm, animal bites, bacteria, toxoplasmosis, and salmonella.

Rabies
<http://www.nlm.nih.gov/medlineplus/rabies.html>

MedlinePlus is a consumer health resource developed by the U.S. National Library of Medicine that provides extensive information about more than 650 diseases and conditions. The information about rabies covers overviews, prevention/screening, organizations, and statistics.

Rabies
<http://www.cdc.gov/ncidod/dvrd/rabies/Ques&Ans/q&a.htm>

The U.S. Centers for Disease Control and Prevention (CDC) is concerned with the monitoring and prevention of disease. The CDC site's information about diseases transmitted from animals to humans includes avoiding contact with wild animals, safeguards for HIV-positive individuals, toxoplasmosis, and more. The information about rabies covers vaccines, animal bites, rabies in animals and humans, travel, and more.

OLD AGE

Behavior Problems in Dogs
<http://www.peteducation.com/article.cfm?cls=2&cat=1551&articleid=616>

Veterinarians Foster and Smith discuss behavior problems common to older dogs including separation anxiety, aggression, inappropriate elimination, fear of noise, increased vocalization, restlessness, and cognitive dysfunction. Each behavior problem is explained and remedies are offered, as available.

Common Diseases in Older Dogs
<http://www.peteducation.com/article.cfm?cls=2&cat=1650&articleid=619>

Veterinarians Foster and Smith's information about diseases that are common in older dogs includes cancer, dental diseases, arthritis, kidney diseases, prostate disease, various eye diseases (cataracts, dry eye), bladder diseases, incontinence, various gastrointestinal diseases, and more.

Geriatric Dogs
<http://www.sniksnak.com/doghealth/geriatric.html>

Pawprints and Purrs, Inc.'s information about aging dogs includes care recommendations (regular vaccinations, grooming, food/water access, symptoms of disease, etc.).

The Senior Dogs Project
<http://www.srdogs.com/>

The Senior Dogs Project is a heroic effort to rescue and place older dogs. The project (1) promotes the adoption of older dogs; (2) provides current information on the special care that older dogs need so that they and their human companions may fully enjoy their golden years; and (3) documents the strong, loving bonds that people have with their older dogs. Site features include compelling stories and images of dogs that have been placed, dogs available in specific parts of the United States, links to purebred rescue groups, health tips for caring for senior dogs, and much more.

PARASITES

Canine Heartworm
<http://www.heartwormsociety.org/CanineHeartwormInfo.htm>

The American Heartworm Society works to "promote effective procedures for the diagnosis, treatment, and prevention of heartworm disease." Site information includes clinical signs, diagnosis, physical examinations, serology (blood tests), treatments, preventive measures, and the difference between heartworm infection in dogs and cats (Figure 5.4).

Dog Owner's Guide—Heartworm Disease
<http://www.canismajor.com/dog/hartworm.html>

Dog Owner's Guide includes "more than 300 pages of features, breed profiles, training tips, health information, and articles about shelters, rescue, dogs, and the law." The section related to heartworm disease includes a description of the *Dirofilaria immitis* parasite, diagnostic tests, treatment, and prevention.

Intestinal Parasites of Dogs and Cats
<http://www.cpvh.com/Articles/24.html>

Claws & Paws Veterinary Hospital in Pearland, Texas explains different types of intestinal parasites that affect cats and dogs. Specific parasites mentioned include hookworms, roundworms, whipworms, tapeworms, Giardiasis, coccidiosis, and toxoplasma. The public health significance of each is also mentioned.

Parasites in Dogs <http://www.peteducation.com/category_summary.cfm?cls=2&cat=1619>

Veterinarians Foster and Smith detail parasites that affect dogs including fleas, ticks, mites, and a variety of worms, how they are transmitted, and methods of controlling them.

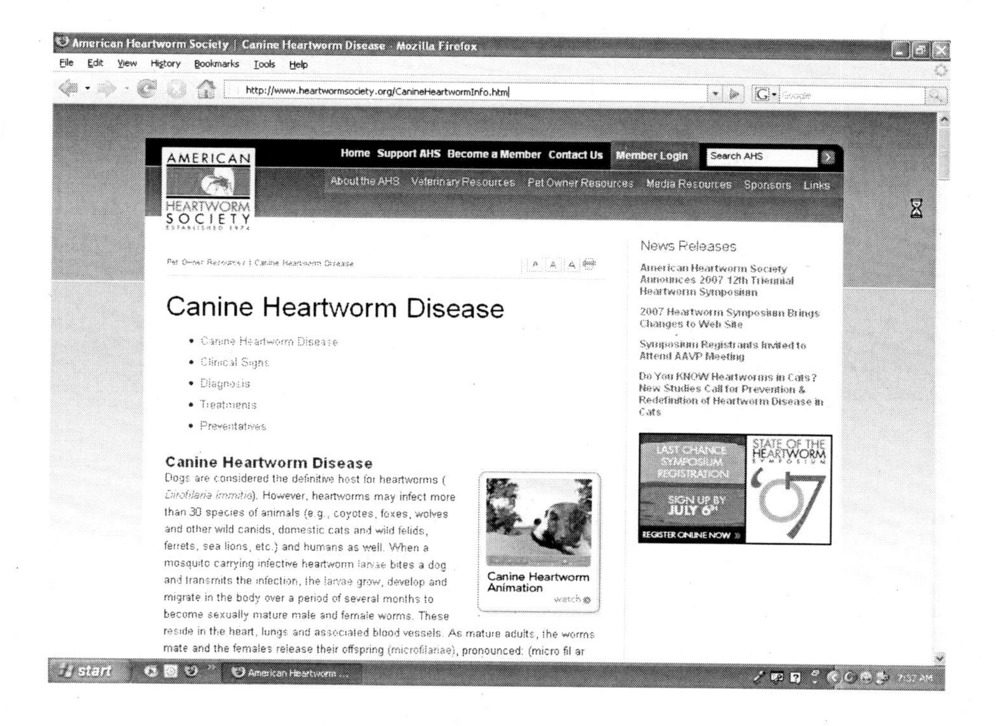

FIGURE 5.4. Canine Heartworm Disease <http://www.heartwormsociety.org/CanineHeartwormInfo.htm> Reprinted with permission.

SKIN, FUR, AND COAT

Causes of Skin Lesions
<http://www.peteducation.com/article.cfm?cls=2&cat=1593&articleid=421>

Veterinarians Foster and Smith explain that hair loss in dogs can be caused by self-licking, allergic reactions, infections (bacterial, parasitic, or fungal), adverse effects of drug therapy, genetics, diabetes mellitus, hormones, and vitamin deficiencies, among other reasons.

Mange
<http://www.canismajor.com/dog/mange1.html>

The Dog Owner's Guide includes a section that describes different mange types (demodectic, cheyletiella, sarcoptic) and their treatment.

Mange—Causes, Symptoms, and Treatment Information
<http://www.animalhospitals-usa.com/dogs/demodectic_parasites_mange.html>

Animal Hospitals USA's site features useful information about illnesses and injuries in "dogs, cats, birds, guinea pigs, gerbils, ferrets, prairie dogs, and reptiles." Demodectic mange afflicts dogs and is caused by the *Demodex canis* mite. Information includes symptoms and treatments.

Mange in Dogs and Cats
<http://www.merckvetmanual.com/mvm/index.jsp?cfile=htm/bc/72005.htm>

Drug manufacturer Merck & Co., Inc.'s veterinary manual includes information about sarcoptic mange in dogs, notoedric mange in cats, otodectic mange in the ears of cats or dogs, cheyletiellosis (walking dandruff), which infects cats, dogs, or rabbits; demodicosis, which afflicts dogs or cats; and trombiculosis, found in cats and dogs. Each type of mange is explained in terms of the parasites involved, diagnosis, and management.

Sarcoptic Mange
<http://www.marvistavet.com/html/sarcoptic_mange.html>

The Mar Vista Animal Medical Center's information about mange explains that this condition is caused by the scabies parasite. This content also includes a description and illustration of the microscopic mite, diagnosis of mange caused by this parasite, and various treatment options.

STOMACH, INTESTINES, AND DIGESTIVE SYSTEM

Anal Sacs
<http://www.marvistavet.com/html/anal_sacs.html>

The Mar Vista Animal Medical Center's site information uses clever animations to show the location and procedure for emptying a dog's anal sacs. Site content explains the significance of "scooting" and the importance of expressing material from anal glands from either cats or dogs.

Gastric Dilatation—Volvulus
<http://www.merckvetmanual.com/mvm/index.jsp?cfile=htm/bc/23305.htm&word=bloat%2cdogs>

Drug manufacturer Merck & Co., Inc.'s veterinary manual explains that gastric bloat (gastric dilatation) is a life-threatening condition that can affect dog breeds with deep chests. Site information covers causes, signs and symptoms, diagnosis, and treatment.

Intestinal Gas—Controlling the Problem
<http://www.peteducation.com/article.cfm?cls=2&cat=1571&articleid=253>

Veterinarians Foster and Smith state that intestinal gas in dogs can be prevented by controlling portions, reducing table scraps, elevating the food dish, exercising after meals, switching to a dog food brand with a lower amount of soy products, and other efforts.

Stomach Bloat in Dogs
<http://www.adoptagolden.com/k9stuff/vetcorner/bloat.htm>

Southeastern Virginia Golden Retriever Rescue Education and Training, Inc. rescues golden retrievers and offers them for adoption or fostering. Veterinarian Anita R. Weidinger defines stomach bloat as when a dog's stomach becomes bloated with air, water, or food.

Vomiting in Dogs
<http://www.sniksnak.com/doghealth/vomiting.html>

Pawprints and Purrs, Inc.'s information explains that vomiting is a symptom, not a disease. This section explains the various reasons why a dog might vomit, and mentions diagnostic tests that can be used.

TEETH AND GUMS (See Also DENTAL CARE)

Canine Dental Care
<http://www.petcaretips.net/canine_dental_care.html>

This information was written by veterinarian Peter Emily and covers treatment and prevention of common problems such as gum disease and tooth fractures in dogs.

Dental Disease in Dogs
<http://www.njvet.com/hlth_dog_dentaldisease.html>

Tenafly & Cliffside Animal Hospitals in New Jersey offer useful health information for cat and dog owners, including this article about tartar buildup, cavities, gum infections, and tooth loss.

Dental Disease in Dogs
<http://www.sniksnak.com/doghealth/dental.html>

Pawprints and Purrs, Inc. site's dental disease information explains how tartar builds up, how cleaning teeth helps prevent other diseases, and the complexities of teeth cleaning at the veterinarian's office.

Periodontal Disease
<http://www.peteducation.com/article.cfm?cls=2&cat=1633&articleid=379>

Veterinarians Foster and Smith's site explains that diseases of the gums and the bones surrounding the teeth are caused by a buildup of dental plaque and tartar. This section details the causes, symptoms, diagnosis, treatment, and prevention.

Periodontitis in Dogs
<http://www.petplace.com/dogs/periodontitis-in-dogs/page1.aspx>

Veterinarian Jon Rappaport's site information about advanced gum disease in dogs includes a disease overview, symptoms, diagnosis, prevention, and treatment.

Chapter 6

For Small Pets Only: Birds, Fish, Reptiles, and Small Mammals

Each bird must sing with his own throat.

Henrik Ibsen

I once had a sparrow alight upon my shoulder for a moment, while I was hoeing in a village garden, and I felt that I was more distinguished by that circumstance than I should have been by any epaulet I could have worn.

Henry David Thoreau

BIRDS

Anatomy, Behavior and Diseases of Caged Birds <http://www.peteducation.com/index.cfm?cls=15>

Veterinarians Foster and Smith developed this excellent resource for pet owners of dogs, cats, fish, birds, ferrets, reptiles, and other small pets (gerbils, hamsters, mice, rats, rabbits, guinea pigs, chinchillas, hedgehogs, sugar gliders, etc.). Site information related to birds includes basic care (airline travel, cage cleaning); behavior and behavioral problems; nutrition and diets; diseases; first aid, emergencies, and poisons.

Internet Guide to Pet Health

doi:10.1300/5675_06

Bird Guide
<http://animal.discovery.com/guides/birds/birds.html?clik=animal_leftnav>

Animal Planet features a bird guide with helpful topics such as choosing a bird, bird care, bird behavior, training, safety (plants to avoid, first aid), and more. Bird diseases include aspergillosis, fatty liver, gout, Newcastle disease, Pacheco's disease, Poxvirus, Psittacine beak and feather disease, Psittacine proventricular dilatation syndrome, and Psittacosis.

Birds
<http://www.vetcentric.com/animalArea.cfm?Section=Birds>

VetCentric is a site that focuses on the health and behavior of dogs, cats, horses, birds, fish, ferrets, rabbits, "pocket pets" (tarantulas, hamsters, sugar gliders, guinea pigs, mice, chinchillas, etc.), reptiles, farm animals, and wildlife. The information specific to pet birds includes biting, bird toys, determining the gender of a bird, West Nile virus, and more.

HotSpot for Birds
<http://www.multiscope.com/hotspot/>

A Los Angeles, California-based bird breeder offers useful information for bird owners including general safety (clipping wings, buying bird toys, smoking by bird owners), nutrition, various diseases (aspergillosis, psittacosis, pseudomonas), safe plants for birds, poisonous/toxic plants, good hygiene, and much more. Site visitors can register for e-mail alerts of new and updated site content.

ThePerch.net
<http://www.theperch.net/>

ThePerch.net is an online bird community that allows parrot fanciers to participate in a threaded discussion group. Other site features include gallery images of various birds, Avian Knowledge Center, which features external links of interest to bird owners, parrot-related products, and more.

FISH

Aquarium Maintenance
<http://www.peteducation.com/index.cfm?cls=16>

Veterinarians Foster and Smith's site information related to aquarium maintenance covers equipment and supplies for salt water and freshwater fish; species of fish and plants; diet and nutrition; and more.

Fish
<http://www.vetcentric.com/animalArea.cfm?Section=Fish>

VetCentric's site explains how to create a healthy habitat for fish, choosing salt water or freshwater fish tanks, vaccinations for fish, and much more.

Fish Guide
<http://animal.discovery.com/guides/fish/fish.html?clik=animal_leftnav>

Animal Planet features a fish guide that is searchable by section (marine aquariums, marine fish, freshwater aquariums, freshwater fish, ponds, frequently asked questions), and covers topics related to choosing a fish, fish care, nutrition, and more.

REPTILES

☑ **Exotic PetVet.Net**
<http://www.exoticpetvet.net/>

The husband and wife team of veterinarian Margaret A. Wissman and zoologist Bill Parsons run Icarus Mobile Veterinary Service and the nonprofit organization, Small World Zoological Gardens in Tampa, Florida. Their Exotic PetVet.Net site features useful information about birds, snakes, lizards, turtles, sugar gliders, rabbits, ferrets, freshwater fish, marmosets, tamarins, and other animals. Other topics include West Nile virus, diabetes in birds, reptile exercise, healthy teeth in bunnies, and more.

☑ Pet Health Information—Reptiles
<http://www.animalhospitals-usa.com/reptiles.html>

Animal Hospitals USA's site features useful information about illnesses and injuries in "dogs, cats, birds, guinea pigs, gerbils, ferrets, prairie dogs, and reptiles." The section about reptiles includes images and text related to iguanas, lizards, snakes, water turtles, box turtles, and desert tortoises. For example, each section covers health, diseases (general, parasitic, infectious) care, feeding, and reproduction (Figure 6.1).

Reptile Guide
<http://animal.discovery.com/guides/reptiles/reptiles.html?clik=animal_leftnav>

Animal Planet features a reptile guide that is organized into sections (body basics and care; iguanas and other lizards; turtles, tortoises, and terrapins; snakes; frogs and amphibians; first aid; safety concerns; and diseases and conditions).

Reptiles Etc.
<http://www.vetcentric.com/animalArea.cfm?Section=Reptiles%20Etc.>

VetCentric's section devoted to reptiles discusses turtles, snakes, frogs, and lizards, and their characteristics and care.

SMALL MAMMALS

This section deals with small mammals such as chinchillas, ferrets, gerbils, guinea pigs, hamsters, hedgehogs, mice, rabbits, rats, and sugar gliders.

Anatomy, Behavior and Diseases of Ferrets
<http://www.peteducation.com/index.cfm?cls=11>

Veterinarians Foster and Smith developed this excellent resource for pet owners of dogs, cats, fish, birds, ferrets, reptiles, and other small pets (gerbils, hamsters, mice, rats, rabbits, guinea pigs, chinchillas, hedgehogs, and

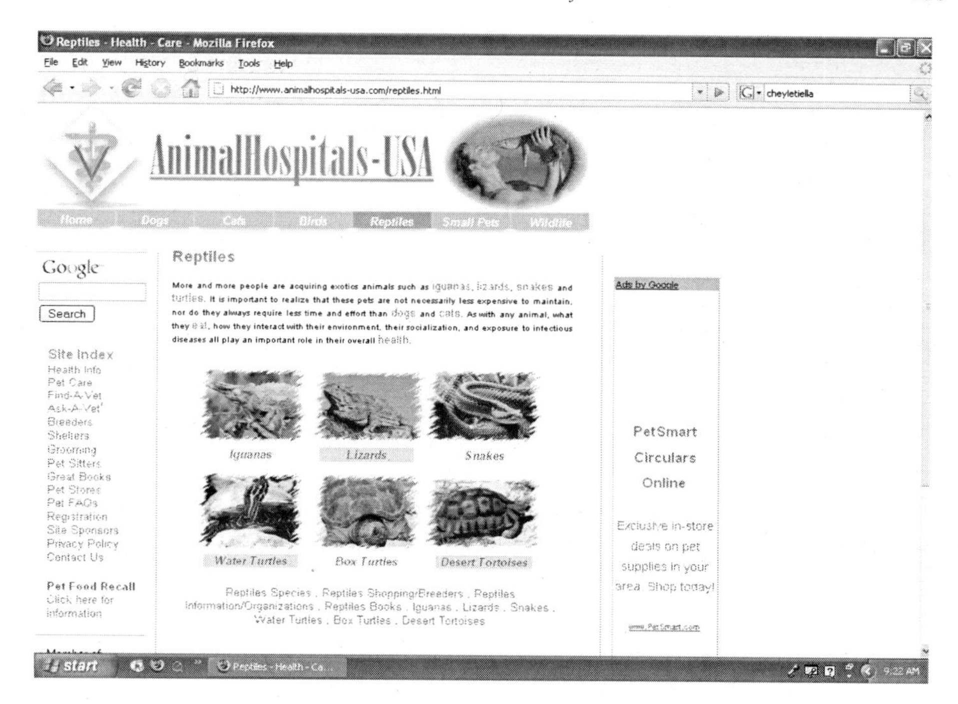

FIGURE 6.1. Pet Health Information—Reptiles <http://www.animalhospitals-usa.com/reptiles.html> Reprinted with permission.

sugar gliders). The section devoted to ferrets covers diseases, parasites, behavior, basic care, diet and nutrition, and emergencies.

☑ Ani-Med—Birds, Fish, Rabbits, Ferrets, Guinea Pigs, Hamsters, Mice, Gerbils, and Rats <http://www.animed.org/index4.htm>

Ani-Med is a site sponsored by the American Society for the Prevention of Cruelty to Animals (ASPCA) that has two different sections: one with information about dogs and cats, and another for birds, fish, rabbits, ferrets, guinea pigs, hamsters, mice, gerbils, and rats. The section about small mammals covers general topics (dental care, grooming, odor control, interactions with children and other pets, and making sure the animal gets enough water). The information specific to individual species deals with understanding the pet, common health problems, proper handling, housing, diet and nutrition, toys, and special treats.

Exotic Pet Care
<http://www.pet-care-plus.com/exotic-pet-care/exotic-pet-care.htm>

Pet Care Plus is a site that focuses on the care and feeding of dogs, cats, birds, and exotic pets, and defines exotic pets as covering "rabbits, guinea pigs, gerbils, hamsters, mice, rats, chinchillas, prairie dogs, hedgehogs, ferrets, pot-bellied pigs, reptiles, amphibians, invertebrates, and fish." Each type of pet features an overview, diet and nutrition needs, health, and housing requirements.

Pocket Pet Primer
<http://www.vetcentric.com/magazine/magazineArticle.cfm?ARTICLEID=1168>

VetCentric's section about pocket pets compares the pets by lifespan, level of interaction with humans, typical behaviors, and drawbacks.

Rabbits—Routine Rabbit Care
<http://www.petwebsite.com/rabbits.asp>

Pet Web Site, based in Oxfordshire, United Kingdom, provides useful information related to cats, dogs, rabbits, gerbils, guinea pigs, horses, mice, and rats. Topics include pet ownership, pet loss, routine care, training, and more. The information about routine rabbit care covers supplies and equipment, feeding, handling, grooming, training, and health.

Rabbits
<http://www.vetcentric.com/animalArea.cfm?Section=Rabbits>

VetCentric's section on rabbits covers choosing a rabbit, grooming tips for Angora rabbits, preventing hair block, spaying/neutering, using a litterbox, and more.

Rats—Routine Care
<http://www.petwebsite.com/rats.asp>

Pet Web Site's information about routine rat care covers supplies and equipment, feeding, handling, grooming, training, and health.

Small Mammal Health Series
<http://www.veterinarypartner.com/Content.plx?P=H&S=5>

The Veterinary Information Network (VIN) developed this site to provide "veterinarians, veterinary students, and the veterinary industry with the information and tools they need to address the demands of modern veterinary practice." The information specific to small pet mammals includes nutritional issues, abscesses in rabbits, Aleutian disease in ferrets, ferret emergencies, care of hedgehogs, nutrient content for snacks, selecting a veterinarian, and more.

Chapter 7

Pet Food Safety

A pet food recall that started in March 2007 has caused widespread fear and consternation among pet owners living in North America, Europe, and other parts of the world. At first, the alerts involved some gravy-based dog and cat food products that had been supplemented with wheat gluten, a concentrated protein derived from wheat flour that was imported from China to Canada. As pets fell ill or died, rat poison was detected in some of the tainted food, but later, melamine, a plastic resin used as a fertilizer, was implicated. It is believed that other additives such as corn, rice, or soy gluten, also used to boost the protein composition of foods, may be similarly contaminated. The recalls later expanded to include a variety of wet and dry foods with glutens possibly contaminated by melamine, cyanuric acid, ammelide, or ammeline. Later concerns focused on these contaminants entering the human food chain after tainted feeds were given to hogs, farmed fish, chickens, and other livestock. As the investigations into these health concerns continue to unfold, readers are advised to track this situation closely by consulting some of the links detailed in this chapter, and to pay close attention to pet food ingredients. Some pet owners have reacted by preparing their own pet foods or switching to more natural products or brands with fewer additives. Readers needing a more complete picture of food safety can consult *Internet Guide to Food Safety and Security,* published by The Haworth Press.

FOOD SAFETY INFORMATION

Additives in Meat and Poultry Products <http://www.fsis.usda.gov/OA/pubs/additive.htm>

The Food Safety Inspection Service (FSIS) is an agency within the U.S. Department of Agriculture that concerns itself with the labeling and packag-

Internet Guide to Pet Health

doi:10.1300/5675_07

ing of "the nation's commercial supply of meat, poultry, and egg products." This document explains how food additives are monitored, regulated, and labeled.

☑ Center for Food Safety and Applied Nutrition (CFSAN) <http://vm.cfsan.fda.gov/>

The U.S. Food and Drug Administration's Center for Food Safety and Applied Nutrition is one of six product centers. Site features of interest to pet owners include food ingredients and packaging, imports, inspections, compliance, enforcements, recalls, and food labeling (Figure 7.1).

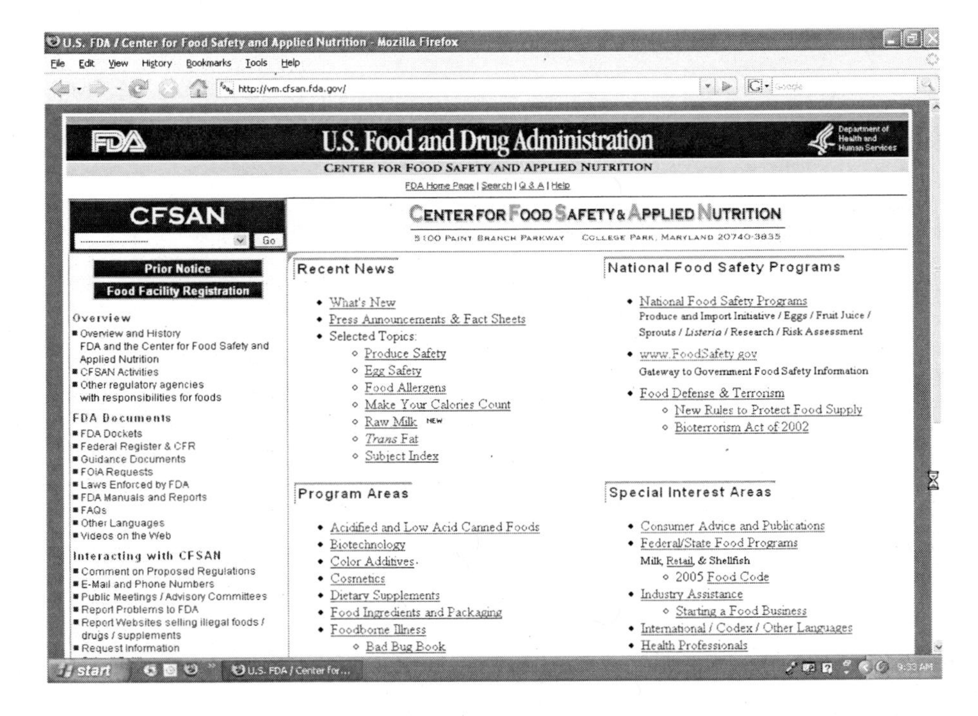

FIGURE 7.1. Center for Food Safety and Applied Nutrition (CFSAN) <http://vm.cfsan.fda.gov/>

☑ Consumer.gov <http://www.consumer.gov/>

This gateway of U.S. government information focuses on several areas of concern to consumers such as food, product safety, transportation, and technology. Site content related to labeling, product recalls, and topics in the spotlight will be of potential interest to persons concerned about the safety of pet foods (Figure 7.2).

☑ Food and Nutrition Information Center (FNIC) <http://fnic.nal.usda.gov/>

The U.S. Department of Agriculture (USDA) and the Agricultural Research Service (ARS) host FNIC in partnership with the University of

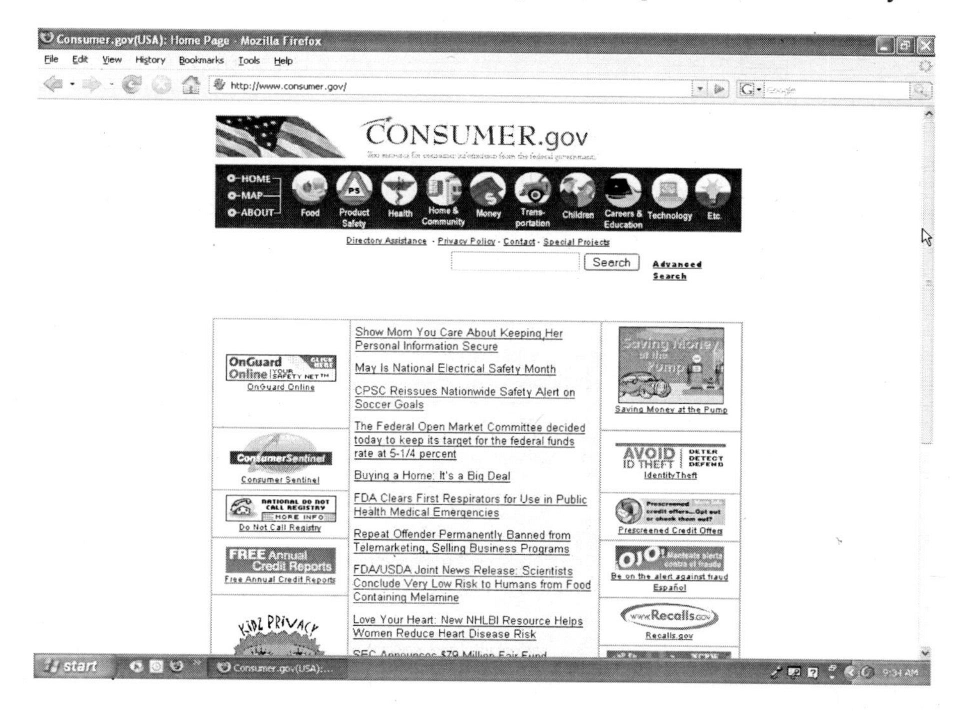

FIGURE 7.2. Consumer.gov <http://www.consumer.gov/>

Maryland and Howard University. Site features specific to food safety include information about food composition, labeling, and supplements, prevention of food borne illnesses, and a discussion group related to food safety (Figure 7.3).

Food Safety and Microbial Hazards
<http://www.foodsafetynetwork.ca/food.htm>

Canada's Food Safety Network's motto is "safe food, from farm to fork." The site's annotated links of potential interest to pet owners include food borne illnesses; food additives, pathogens, and production; and food codes and regulations.

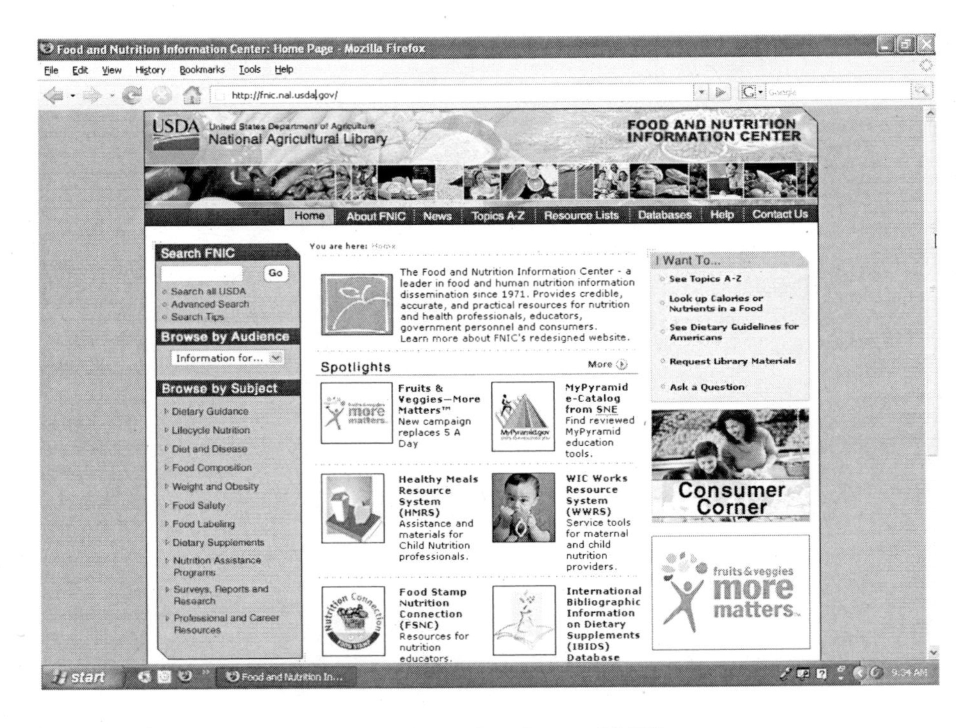

FIGURE 7.3. Food and Nutrition Information Center (FNIC)
<http://fnic.nal.usda.gov/>

☑ Food Safety Risk Analysis Clearinghouse
<http://www.foodriskclearinghouse.umd.edu/>

The Joint Institute for Food Safety and Applied Nutrition (JIFSAN) is a collaborative effort of the U.S. Food and Drug Administration (FDA) and the University of Maryland. JIFSAN's Food Safety Risk Analysis Clearing house site includes categories related to risk analysis, risk assessments, risk profiles, nutrition and labeling, hot topics, and more. Recent hot topics have included acrylamide, dioxin, and bovine spongiform encephalopathy.

Food Safety Throughout the Food System
<http://foodsafety.cas.psu.edu/>

Pennsylvania State University's College of Agricultural Sciences hosts this food safety site. Site features include a Food Safety Database, news updates, rumor control, information about recalls, discussion groups, and links to various publications.

FoodSafety.gov
<http://www.foodsafety.gov/>

The FoodSafety.gov site describes itself as a "gateway to government food safety information." Site features include safety alerts from various government agencies, consumer advice such as safe handling of foods and reporting food borne illnesses, information about food borne pathogens from local, regional, and federal sources, and much more.

☑ MedlinePlus—Food Contamination and Poisoning
<http://www.nlm.nih.gov/medlineplus/foodcontaminationandpoisoning.html>

MedlinePlus organizes authoritative content and links including an overview and latest news about contaminated foods, research, diagnosis/symptoms of specific conditions, dictionaries/glossaries, and subject-related links. The information specific to pet food recalls includes updates on tainted animal feed, risks to humans, and related topics (Figure 7.4).

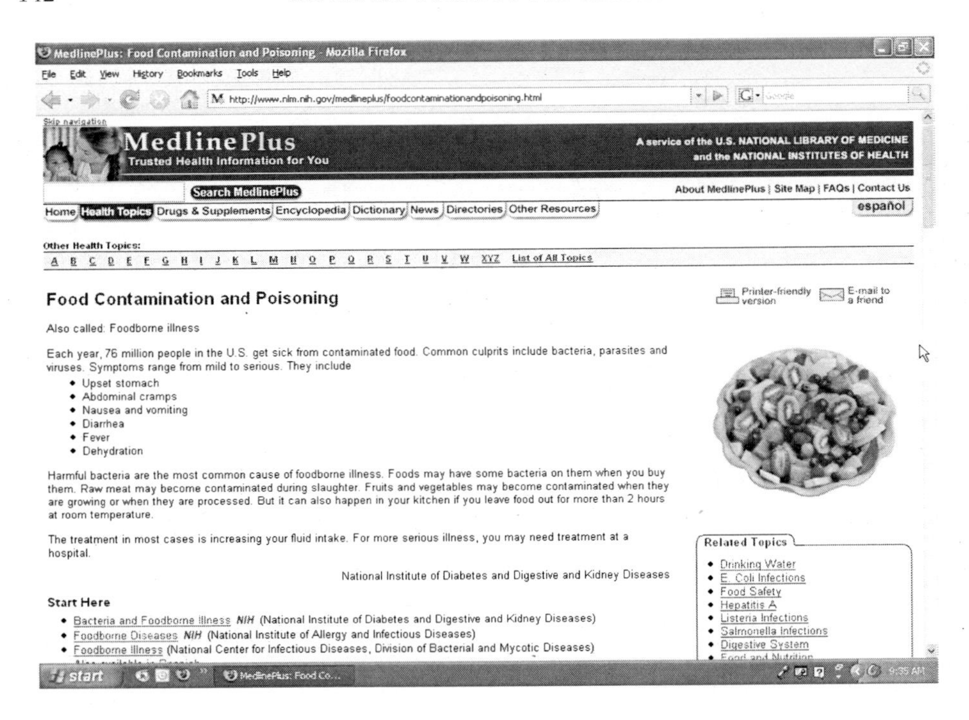

FIGURE 7.4. MedlinePlus—Food Contamination and Poisoning <http://www.nlm.nih.gov/medlineplus/foodcontaminationandpoisoning.html>

RECALLS

☑ Food and Drug Administration (FDA) <http://www.fda.gov/>

The U.S. Food and Drug Administration (FDA) is an agency within the Department of Health and Human Services. Site features include news, hot topics, specific agency activities, newsletters sent by e-mail or RSS, and publications including *FDA Consumer.* The Animal Feed and Drugs section features a Pet Food Recall page (Figure 7.5).

FIGURE 7.5. Food and Drug Administration (FDA) Pet Food Recall/Tainted Animal Feed
<http://www.fda.gov/oc/opacom/hottopics/petfood.html/>

☑ FSIS Recalls <http://www.fsis.usda.gov/Fsis_Recalls/index.asp>

The U.S. Department of Agriculture's Food Safety Inspection Service (FSIS) site covers definitions, frequently asked questions, procedures, and archived cases of contaminated, adulterated, or misbranded foods related to food recalls (Figure 7.6).

Howl 911's Pet Food Recall <http://www.howl911.com/>

This site was developed in March 2007 soon after the first pet food recalls, to consolidate information about commercial and private label brands

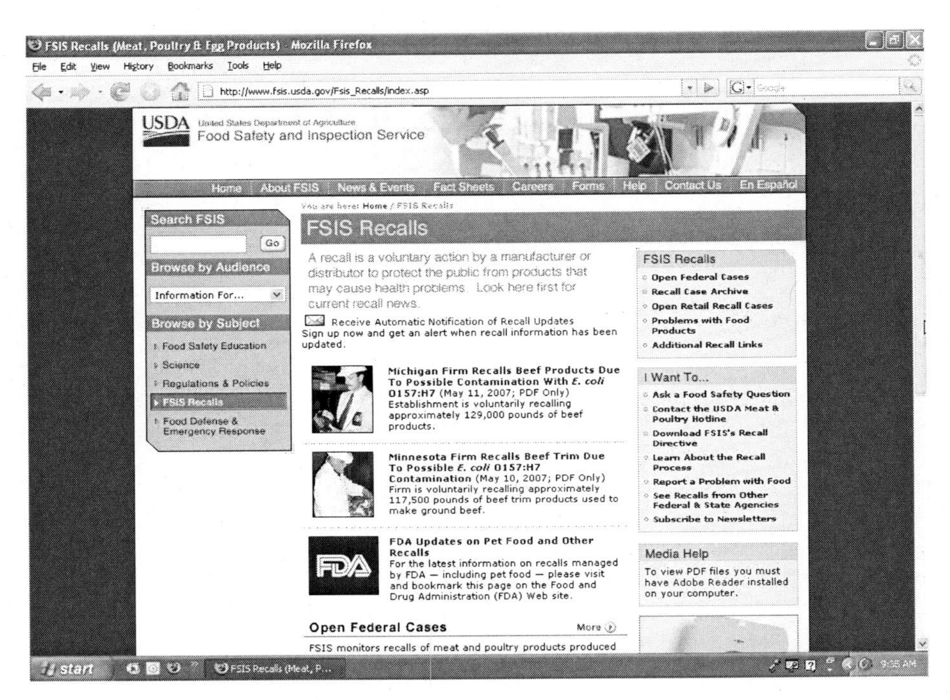

FIGURE 7.6. FSIS Recalls
<http://www.fsis.usda.gov/Fsis_Recalls/index.asp>

of pet food affected by the recalls. Site information includes lists of recalled brands, manufacturers of specific brands, current and archived news headlines, discussion forums, blogs with related information, and more.

Index of Canadian Food Recalls <http://www.inspection.gc.ca/english/corpaffr/recarapp/recaltoce.shtml>

The Canadian Food Inspection Agency provides e-mail notification of various food recalls and allergy alerts. Site information can be sorted by type of food commodity (dairy, eggs, feeds, fertilizers, fish and seafood, fresh fruits and vegetables, grains, honey, potatoes, etc.) or key topic (animal health, biotechnology, horticulture, food recalls, processed foods, etc.). Information specific to melamine-contaminated pet foods refers to import controls and testing of "wheat, rice, soy, and corn gluten."

Pet Food Recall (AVMA)
<http://www.avma.org/aa/petfoodrecall/>

The American Veterinary Medical Association (AVMA) developed this list of pet food recall information for pet owners and veterinarians. The information includes recalled products, pet food manufacturers, news releases, and procedures for submitting samples and/or reported cases of affected animals.

Pet Food Recall (Cornell College of Veterinary Medicine)
<http://www.vet.cornell.edu/news/FoodRecall/>

Cornell University's College of Veterinary Medicine site developed a pet food recall page that features answers to frequently asked questions, recent press releases, information for pet owners, and information for veterinarians.

Recalls and Product Safety News—Consumer Product Safety Commission
<http://www.cpsc.gov/cpscpub/prerel/prerel.html>

Although the Consumer Product Safety Commission is a U.S.-based agency, which focuses on the safety of durable consumer goods such as appliances, household wares, and the like, the site includes recall, product safety, and poison prevention information of interest and value to pet owners. The recalls and product safety news database is searchable by date, product type, company, product category, and press release number.

Urgent Bulletins—University of Guelph
<http://www.labservices.uoguelph.ca/urgent.cfm>

The University of Guelph's Laboratory Services Division developed an urgent bulletin on pet food recalls that includes compelling photographs of contaminated wheat gluten, microphotographs of crystallized urine samples, spectrographic images of urinary crystals from affected cats, and information for "corporate clients or veterinarians" wishing to submit food samples for testing.

NATURAL PET FOODS

Alternatives to Conventional Commercial Pet Foods
<http://www.vet.upenn.edu/nutrition/altdiets.htm>

The University of Pennsylvania's School of Veterinary Medicine site developed this handy comparison chart of the pros and cons of using home-prepared, organic, vegetarian, or raw food diets to feed pet animals.

Dog Food Recipes
<http://www.i-love-dogs.com/dog-food-recipes.html>

This resource is an online cookbook of recipes including treats, snacks, casseroles, and stews for dogs.

Making Cat Food Yourself
<http://www.catnutrition.org/foodmaking.html>

Catnutrition.org stresses the importance of balanced nutrition for kittens and cats and understanding cat food ingredient labels. Site information includes recipes, equipment, food supplements, storage tips, and more.

Natural and Healthy Pet Food Recipes
<http://www.e-healthypetfood.com/>

This site features a variety of healthy and tantalizing meat and vegetable dishes for dogs, cats, rabbits, and reptiles (lizards, turtles, iguanas, etc.). Examples include vegetarian chowder, flea-fighting dog biscuits, cabbage rolls, tofu and liver loaves, tacos, chicken and rice, and much more.

Natural Dog Food Web Site
<http://www.naturaldogfood.com/>

Carol Brescher Boyle, author of *Natural Food Recipes for Healthy Dogs* developed this site as an emergency measure for pet owners concerned about possible pet food contamination. Site features include excerpts from her book, dog food recipes, suggested reading, and more.

Pet Food Recipes
<http://www.recipezaar.com/recipes/pet-food>

Despite the name of this Recipezaar section, the recipes are for dog treats (biscuits, scones, and cookies) and meaty dog meals. Each recipe includes easy-to-follow instructions and detailed nutritional information per serving (calories, fat content, sodium, sugars, protein, etc.).

Pet Food Recipes—CooksRecipes
<http://www.cooksrecipes.com/category/pet-recipes.html>

CooksRecipes.com is a site that focuses on cooking, including a batch of recipes for making food for cats or dogs. Recipes feature tuna, cheddar cheese, bacon, oatmeal, peanut butter, chicken liver, and other savory ingredients.

PetDIETS.com
<http://www.petdiets.com/>

PetDIETS features content developed by veterinary nutritionists including nutritional needs, energy requirements, dangers of homemade diets, understanding food labels, and articles that include recipes that are downloadable for a fee.

Rawfood Diet for Dogs and Cats
<http://www.shirleys-wellness-cafe.com/sampleraw.htm>

Shirley's Wellness Café focuses on the importance of raw foods for dogs and cats. The site includes a raw foods calculator; recipes for dogs, cats, and other animals; and the health benefits of raw ingredients rather than cooked foods.

Tips N Tricks Pet Food Recipes
<http://www.budget101.com/pets.htm>

Budget101.com is a site that focuses on frugal living. Information about pet food includes several economical recipes for kittens, cats, dogs, birds, and horses, and other recipes for repelling fleas, ticks, and deer.

TreatWorld
<http://www.treatworld.com/>

TreatWorld includes a modest collection of recipes for preparing treats and meals for puppies and dogs made out of pumpkin, yogurt, carrots, peanut butter, bacon, and other appealing and healthy ingredients.

Glossary

Some of the terms used in this guide may be unfamiliar to the average person. These definitions and information about acronyms were derived from several excellent sources including:

- **Medical Glossary.org**
  <http://www.medicalglossary.org/>
- **MedlinePlus Merriam-Webster Medical Dictionary**
  <http://www.nlm.nih.gov/medlineplus/mplusdictionary.html>
- **OneLook Dictionary Search**
  <http://www.onelook.com/>
- **Acronym Finder**
  <http://www.acronymfinder.com/>

acronyms: Words formed from the first letters of two or more words in a name or phrase, such as SPCA for Society for the Prevention of Cruelty to Animals.

aggregated: Collected or combined. In the context of health information, this term is used to indicate grouped resources that can be searched with a common interface.

AGRICOLA: An article citation database that indexes journal articles, book chapters, short reports, and reprints.

allergens: Substances that cause allergic reactions. *See also* ALLERGY.

allergy: High sensitivity to irritants such as pollens, foods, and micro-organisms, resulting in physical reactions.

alopecia: Hair loss.

Internet Guide to Pet Health

doi:10.1300/5675_08

amebiasis: Infection or disease caused by amebas, protozoan organisms that can cause diarrheal illness. Amebiasis is often caused by the *Entamoeba histolytica* organism.

amphibians: Animals that function both on land and in water such as frogs, toads, and salamanders.

antigens: Substances used to stimulate an immune response.

antimicrobial resistance: Microorganisms that can modify themselves in ways that reduce the effectiveness of drug treatment.

antimicrobials: Substances that kill or reduce infectious agents.

arthropods: A group of animals including crustaceans (crabs, shrimps, lobsters), spiders, mites, centipedes, and insects.

atopic: Allergic.

atrophy: Reduction in size or wasting of animal tissue or organs.

authoritative: As related to Web site content, accurate and reliable.

avian: Related to birds.

bacteria: Single-celled organisms that live independently or feed parasitically off other organisms. Bacteria can be harmless or harmful.

bibliographic: As related to informational databases, citations that include author, title, source, and abstract, elements that are needed for finding full-text information in print or electronic publications.

bird influenza: A contagious respiratory tract infection that affects birds.

blog: Web + log = blog. Blogs serve as online diaries or journals.

bovine: Related to cattle.

bovine spongiform encephalopathy: A neurodegenerative disease in cattle that is caused by a prion that has been transmitted by feed infected with animal tissue; also called mad cow disease.

Brucella abortus: A Gram-negative bacterium that causes brucellosis.

brucellosis: A bacterial disease that affects animals and humans and is caused by the *Brucella abortus* bacterium.

calculi: Stones.

Campylobacter: A type of Gram-negative bacteria. Campylobacteriosis refers to the infection by this organism.

canines: Animals in the dog family.

cardiomyopathy: Enlargement or thickening of the heart muscle which results in diminished or abnormal function.

CJD: *See* CREUTZFELDT-JAKOB DISEASE.

clinical trials: Series of treatments used to evaluate the effectiveness of specific medications or medical procedures.

cloning: The process of growing a cell or organism that is genetically identical to the original.

coccidia: A single-celled parasite that causes diarrhea and intestinal damage.

communicable diseases: Diseases that are contagious or easily transmitted from one organism to another.

Creutzfeldt-Jakob Disease (CJD): An incurable neurodegenerative disease.

cryptococcosis: A fungal disease caused by *Cryptococcus neoformans,* a yeast.

cryptosporidiosis: A protozoal disease caused by *Cryptosporidium parvum.*

dander: Particles that are shed from skin, feathers, or hair.

degenerative: Deteriorating or showing loss of function.

dermatitis: Skin inflammation.

diagnosis: Methods used to identify characteristics, signs, or symptoms of a disease, condition, or ailment that distinguishes the disease from other diseases.

disease outbreaks: Sudden increases or eruptions of disease activity.

disease surveillance: Close observation and tracking of disease activity by public health authorities. *See also* SURVEILLANCE.

disease transmission: Transfer of disease to another organism.

disease vectors: Organisms such as mosquitoes, ticks, or other organisms that carry disease-causing microorganisms from one host (person or animal) to another person or animal.

dysplasia: Abnormal growth or development of tissues or organs.

eczema: Scaly skin.

elixirs: Substances or tonics, usually sweet or otherwise palatable, intended to cure a variety of conditions and diseases.

emesis: Vomiting.

encephalopathy: A degenerative disease of the brain.

endemic: Diseases common or peculiar to a specific geographic region. Compare this term with epidemic.

epidemic: The outbreak and spread of disease in a geographic area. Compare this term with endemic.

epilepsy: A brain disorder that is characterized by seizures or sudden involuntary movements.

epizootic: A disease that affects many organisms at the same time within a specific region or geographic area.

Escherichia coli: Known more familiarly as *E. coli,* this organism exists normally within the gastrointestinal system but dangerous forms are responsible for serious blood or intestinal diseases.

etiology: The study of the causes of disease.

euthanasia: The practice of killing gravely ill, injured, or dying persons or animals.

ezine or **e-zine:** Electronic magazine.

felines: Animals in the cat family.

feral: Previously domesticated and now living in the wild.

fistulae: Abnormal tracts or passages between two organs.

forebrain: The front part of the brain that is responsible for memory, motor functions, thinking, and emotions.

fungal: Relating to a fungus or fungi.

fungi: Plural of fungus; plant organisms that include yeasts, molds, mildews, mosses, algaes, smuts, and mushrooms.

giardiasis: A protozoal disease that can cause stomach upset and diarrhea.

gingivitis: Inflammation of the gum tissue around the teeth.

hantaviruses: Viruses transmitted by rodents.

heartworm: A parasitic worm transmitted by mosquitoes that invades the pulmonary arteries.

hematemesis: Vomiting blood.

hemorrhagic: Characterized by sudden and/or excessive loss of blood.

hepatitis: Inflammation of the liver caused by infection or toxins.

hormones: Substances that affect growth or metabolism.

host: An organism on which another organism (parasite) feeds or lives.

humane: Concerned with the alleviation or prevention of suffering.

husbandry: The practice of breeding animals.

hypertext: Text that links to other text.

immune response: A physical response to an antigen by specific antibodies.

immunization: A procedure that introduces specific antigens to induce an immune response, in an effort to improve the ability to resist infection, and to render immunity for a specific disease. This term is used interchangeably with vaccination.

immunocompromised: Lack of a normal immune response, as a result of disease, malnutrition, or immunosuppressive therapy.

immunosuppression: Suppression of the immune response by disease, medications or radiation.

insecticides: Chemicals used to kill insects. *See also* PESTICIDES.

ischemia: Decreased blood supply.

leishmaniasis: A protozoal disease spread by sandflies.

MEDLINE: A bibliographic database of the published biomedical literature, developed by the U.S. National Library of Medicine (NLM).

MedlinePlus: A consumer health resource developed by the U.S. National Library of Medicine that provides extensive information about more than

650 diseases and conditions. Also includes lists of hospitals and physicians, medical encyclopedia, medical dictionary, information about prescription and nonprescription drugs, and health information from the media.

metabolism: Chemical processes within a living organism.

microorganisms: Tiny life forms such as bacteria and protozoa that can be harmful or beneficial.

mycobacteria: A type of bacteria, some of which cause tuberculosis and leprosy.

mycotic: Infections or diseases caused by a fungus or fungi.

mycotoxins: Toxins produced by a fungus or fungi.

nematodes: Roundworms.

nephrology: A specialized field of medicine that studies the causes, diagnosis, and treatment of kidney diseases.

neutering: Sexual sterilization of a male animal.

notifiable diseases: Diseases required by law to be reported to public heath authorities.

nutrients: Substances such as trace elements that promote growth.

occurrence: Instances of a specific disease appearing in a population.

oncology: A specialized field of medicine that studies the causes, diagnosis, and treatment of cancer.

organic: Derived from living organisms. Compare this term with inorganic.

outbreaks: *See* DISEASE OUTBREAKS.

palatable: Good tasting.

parasites: Organisms that feed or live on another organism, often called the host.

parvo: Parvoviruses cause highly contagious diseases in dogs and cattle, and are spread by infected feces.

patella: The bony part of the knee or kneecap.

peer-reviewed: Professional evaluation of a colleague's work. This term usually refers to the evaluation of articles for publication or grants for funding.

periodontics: A field of dentistry that treats gum and bone diseases.

peritonitis: Inflammation of the abdominal cavity lining.

pesticides: Chemicals used to kill pests, especially insects, worms, and other organisms.

pica: The tendency to eat items with no known nutritional value, which are potentially harmful, such as paint chips, dirt, and other materials.

portals: As related to Web sites, gateways to resources..

primer: An introductory textbook or guide.

prions: Protein particles similar to viruses thought to cause Creutzfeldt-Jakob Disease and other degenerative diseases of the nervous system. The term is derived from the combination of the words proteinaceous + infectious = prion.

prognosis: Predicted survival or recovery.

protozoa: Single-celled organisms including amebas, ciliates, flagellates, and sporozoans.

reportable diseases: Diseases required by law to be reported to public heath authorities.

quarantine: Isolation of an individual or animal to prevent the spread of disease.

rabies: A viral infection that attacks the nervous system.

reptiles: A class of vertebrates that includes turtles, tortoises, snakes, lizards, alligators, and crocodiles.

renal: Related to the kidney.

residue: In this context, the amount of a toxic substance that remains on surfaces, in the soil, or in plant or animal tissues.

rodents: Family of animals that includes squirrels, rats, mice, beavers, and chipmunks.

RSS feeds: "Really simple syndication" uses XML technology as a means to distribute content from one site or source to other sites. For example, RSS can be used to feed continually updated headlines from the CNN site to another site.

sarcoptic: Refers to infection with *Sarcoptes scabiei,* an insect parasite.

search directory: A Web tool that organizes Internet resources by subject and/or file type (documents, images, news group messages).

search engine: A Web tool that finds Internet resources based on keywords or phrases typed into a search interface.

spaying: Sexual sterilization of a female animal by removing the ovaries.

species: Category of related organisms that are capable of breeding with one another.

spongiform: With a soft, porous, and sponge-like texture. *See also* BOVINE SPONGIFORM ENCEPHALOPATHY.

stomatitis: Mouth ulcers or sores.

subsets: Parts of a larger collection.

surveillance: The close observation of persons or situations. *See also* DISEASE SURVEILLANCE.

sustainability: Practices used to maintain a healthy environment, particularly recycling.

tachycardia: Rapid heart beat.

terrapins: Turtles.

tickborne: Diseases carried or transmitted by ticks.

tissues: As related to the body, a collection of similar cells that act together and perform specific functions.

toxicants: Poisons or poisonous substances.

toxicology: The study of poisons and their effects on the body.

toxins: Harmful substances that can cause disease.

toxoplasmosis: A parasitic disease caused by *Toxoplasma gondii,* spread by uncooked meats and cat feces, that can cause brain diseases and birth defects.

transmission: Transfer of a disease from one person or animal to another.

tularemia: A bacterial disease transmitted by insect bites which affects humans and animals.

urology: A specialized field of medicine that studies the causes, diagnosis, and treatment of diseases involving the bladder, kidneys, and urinary tract.

uveitis: Inflammation of eye membranes.

vaccination: Introduction of harmless or killed bacterial or viral organisms to render immunity for a specific disease; a procedure that introduces specific antigens to induce an immune response, thereby improving the ability to resist infection. This term can be used interchangeably with immunization.

vaccines: Preparations of microorganisms used to stimulate an immune response to prevent future infection with similar microorganisms.

vaccinosis: Disease caused by vaccines.

vectors: *See* DISEASE VECTORS.

vestibular: Related to the inner ear.

veterinary medicine: A specialized field that focuses on the causes, diagnoses, and treatment of diseases in animals.

viral: Relating to or caused by a virus.

viruses: Tiny parasites that feed on plants, animals, and bacteria, often causing diseases.

Web browser: A software program (such as Netscape, Internet Explorer, or Firefox) that is used to view content (text, images) on the World Wide Web.

white paper: An authoritative report or position paper on an issue of concern or controversy.

XML: EXtensible Markup Language that is used to design and display Web content.

zoonosis: Transmission of disease from animals to humans. The plural form of this term is zoonoses.

Bibliography

Alderton, D. (2001). *The Ultimate Encyclopedia of Small Pets and Petcare: The Essential Family Reference Guide to Caring for the Most Popular Pet Species and Breeds, Including Small Mammals, Bird, Reptiles, Invertebrates, and Fish*. London: Lorenz.

Becker, M. (2002). *The Healing Power of Pets: Harnessing the Amazing Ability of Pets to Make and Keep People Happy and Healthy*. New York: Hyperion Books.

Carroll, D. (2001). *The ASPCA Complete Guide to Pet Care*. New York: Plume.

Mammato, B. (1997). *Pet First Aid: Cats and Dogs*. St. Louis, MO: Mosby.

Marder, A. (1994). *Your Healthy Pet: A Practical Guide to Choosing and Raising Happier, Healthier Dogs and Cats*. Emmaus, PA: Rodale Press.

Pinney, C. S. (2004). *The Complete Home Veterinary Guide*. New York: McGraw-Hill.

Podberscek, A. L., and Paul, E. S. (2000). *Companion Animals and Us: Exploring the Relationships Between People and Pets*. Cambridge: Cambridge University Press.

Puotinen, C. J. (2001). *The Encyclopedia of Natural Pet Care*. New York: Plume.

Internet Guide to Pet Health

doi:10.1300/5675_09

Index

Page numbers followed by "f" indicate a figure.

doi:10.1300/5675_10